MOUNT MARY COLLEGE LIBRARY
Milwaukee, Wisconsin 53222

PATTERNS OF SYMMETRY

University of Massachusetts Press, Amherst   1977

Edited by
Marjorie Senechal
and
George Fleck

PATTERNS OF SYMMETRY

To the memory of Dorothy Wrinch, 1894–1976

Copyright © 1974 by Marjorie Senechal, George Fleck,
and Allan Ludman.
New material copyright © 1977 by Marjorie Senechal
and George Fleck
All rights reserved
Library of Congress Catalog Card Number 76–56775
ISBN 0–87023–345–9
Printed in the United States of America
Designed by Mary Mendell
Library of Congress Cataloging in Publication Data
appear on the last printed page of this book.
The support of Smith College is gratefully acknowledged.
"Notes of an Alchemist" is reprinted by permission of
Charles Scribner's Sons from *Notes of an Alchemist* by
Loren Eiseley. Copyright © 1972 Loren Eiseley.

Contents

vii Preface

1. Introduction

3 Patterns of Symmetry: Seek and Ye Shall Find

2. Proceedings of the Symmetry Festival

23 The Many Facets of Symmetry
ARTHUR LOEB 28 Color Symmetry and Its Significance for Science
ALICE DICKINSON 44 Change Ringing: Theory and Practice
ELLIOT OFFNER 50 Renaissance Typographic Ornament: Origins, Use, and Experiments

3. Reflections

SIR JOHN DAVIES 76 Orchestra, or a Poem on Dancing
GEORGE FLECK 79 Symmetry: Making the Unseeable Imaginable
ALLAN LUDMAN 87 Symmetry: The Framework of the Earth
MARJORIE SENECHAL 91 Symmetry: The Perception of Order
PHILIP REID 96 Symmetries in Plants
GEORGE FAYEN 104 Ambiguities in Symmetry-Seeking: Borges and Others
LOREN EISELEY 118 Notes of an Alchemist

4. A Final Focus

MARJORIE SENECHAL 123 The Theory of Patterns
GEORGE FLECK 142 A Final Focus

147 Selected Bibliography
151 Notes

Preface

In February 1973 the world's first Symmetry Festival was held at Smith College, Northampton, Massachusetts. This book has its origins in that unique event. We have tried to convey the spirit, scope, and substance of the festival in a volume that would not only be informative and challenging to all who are fascinated by pattern and form in any field, but would also be useful to students and teachers in a wide variety of subjects and at different levels.

The patterns of symmetry discussed here extend beyond the visible symmetries of objects such as butterflies, furniture, and plants to include some of the less exact—but no less significant—symmetries of art, music, dance, and literature. The patterns also extend beneath the visible level to the symmetries of processes by which some objects are constructed. We hope that you will be as fascinated as we are by the discovery of patterns of symmetry throughout the humanities and science, and will come to appreciate the challenge implicit in the Principle of Symmetry, first stated about 1890 by Pierre Curie:

> When certain causes produce certain effects, the elements of symmetry in the causes ought to reappear in the effects produced.
>
> When certain effects reveal a certain dissymmetry, this dissymmetry should be apparent in the causes which have given them birth.
>
> The converse of these two statements does not hold, at least practically; that is to say, the effects produced can be more symmetrical than their causes.

We are pleased to include contributions from six of our colleagues. Alice Dickinson is professor of mathematics at Smith College. She introduced change ringing to the campus, first through hand bells and later through the eight-bell tower at the Mendenhall Center for the Performing Arts. George Fayen was associate professor of English at Smith College until 1976. "Ambiguities in Symmetry-Seeking" was written especially for this volume. Arthur Loeb is senior lecturer in the Department of Visual and Environmental Studies at Harvard University. His many vocations include crystallography and music. Allan Ludman was assistant professor of geology at Smith College until 1975. Together with the editors, he helped to plan and direct the Symmetry Festival in 1973. Elliot Offner is Andrew Mellon Professor of Art at Smith College and Printer to the

College. He was a founder of the Rosemary Press in Northampton. Philip Reid, a botanist, is associate professor in the biological sciences at Smith College. "Symmetries in Plants" was written especially for this volume.

We would like to express our gratitude to Mary McDougle, secretary of Smith College and of its Lecture Committee, for her encouragement and support at every stage of the Symmetry Festival and in the preparation of *Patterns of Symmetry*. We are further indebted to Marshall Schalk, Professor Emeritus of Geology, for his photography, and to Julia Britt for typing several versions of the manuscript.

The preparation and publication of this book was supported in part by grants from the Sarah Sanderson Vanderbilt Memorial Fund and from the Edwin H. and Helen M. Land Fund of Smith College.

Finally we would like to express our gratitude to the late Dorothy Wrinch. Dr. Wrinch held various positions as lecturer and research professor of physics at Smith College from 1941 to 1971. In informal discussions and in courses such as "Form and Structure in Nature," "The Structure of Large Molecules," and "Physical Crystallography," she emphasized the centrality of the principle of symmetry and the thesis that no phenomenon can be fully understood until the relationship between the symmetry of the causes and the symmetry of the results has been elucidated. Her lifelong research into the structure of proteins was guided by this mandate. She brought her knowledge of mathematics, physics, chemistry, and biology to bear on this problem, and her technical work was always enriched by her understanding of the roles of symmetry in other fields. When we came to teach at Smith we found a friend and teacher in her, and our approaches to symmetry were developed under her influence. Thus in a sense the Symmetry Festival was a reflection of Dorothy Wrinch; this book, a reflection of the festival, is dedicated to her memory.

M.S.
G.F.
Northampton, Massachusetts
July 1976

1 INTRODUCTION

Patterns of Symmetry: Seek and Ye Shall Find

Symmetry can be seen just about everywhere when you know how to look for it. As a broad notion of balance and harmony, symmetry has been familiar to most peoples since the earliest times. In geometrical patterns from all ages we can readily see how artists and artisans have worked creatively within the tight constraints imposed by strictly interpreted symmetry principles. More liberally construed symmetry ideas are often found in paintings, sculpture, music, and poetry. Symmetry operations are frequently the rules by which dancers move; symmetrical motions constitute the dance. We also find that the language of symmetry is often well-suited for discussing artistic works, even when the significant features appear to be deviations from symmetry—or deliberate attempts to exclude it.

Just as significant for human culture are the patterns and balances of cause and effect, which we seek to understand through philosophy and science. Here symmetry-based theories may be conspicuous, the uses of symmetry quite elaborate and detailed. "As far as I can see," wrote Weyl[1], "all *a priori* statements in physics have their origin in symmetry." Symmetry underlies many areas of perception and expression, finding its application and elucidation through the particular concepts and tools of those fields. But in addition to specialized uses, symmetry principles can also serve to unify large bodies of knowledge.

Only rarely is an individual who works with symmetry fully aware of the extent of its manifestations. One's own jargon can obscure the role that symmetry plays, and other people's ways of dealing with their technical matters can mask what may well be a common enterprise. Artists, musicians, and scientists deal with similar concepts but often are prevented by their different languages from appreciating the extent to which they are engaged in but different aspects of the same work. Perhaps it has always been this way, although our own highly specialized time seems to have intensified feelings of separateness.

This volume, *Patterns of Symmetry*, had its origins in a guidebook that we prepared for the Symmetry Festival held at Smith College in February 1973. In revising and expanding the booklet, we have brought together the "Proceedings of the Symmetry Festival" (insofar as it is possible to reproduce those proceedings on the printed page) and have also included additional material under the general heading "Reflections." We hope to demonstrate, through the varied expositions and juxtapositions, that the many facets of symmetry are united in a symmetric whole.

On the next few pages, we introduce the basic concepts of classical symmetry, pointing out some of the ways in which these concepts have been extended in recent years. We also discuss some of the ways in which the contributions to the "Proceedings" illustrate these concepts.

Though symmetry is familiar to us all, the concept must be formulated in a precise way before its properties can be defined and analyzed, and the relations among its properties elucidated. If you are already acquainted with symmetry theory, you may wish to turn directly to Part 2 to discover patterns of symmetry for yourself. But if you are a newcomer to the subject, we hope that what we have to say here will help you recognize symmetries and their relationships. Therein lies the special joy and intellectual excitement of recognizing a pattern, of perceiving the relation of the parts to the whole.

Reflection

Reflection is the symmetry that is best known and most frequently found in nature. A mirror reproduces exactly what is "sees," but reverses the spatial order: the mirror image of a right hand is a left hand, because the arrangement of fingers is reversed. Extending this analogy, we say that anything which can be thought of as divisible into two equivalent but mirror-image halves has mirror symmetry. The two halves are reflections of each other and the hypothetical plane that separates the two halves is called a mirror plane, or just a mirror. (This ideal mirror, like a real mirror, has two sides. Unlike a real one, both of its sides are reflective. It is transparent and has no thickness.)

Using the language of symmetry theory, we say that the abstract mirror plane is a *symmetry element*. The corresponding action (reflection in that plane) is called a *symmetry operation*. The symmetry of any object can always be described in terms of its symmetry elements; it can also be described in terms of its symmetry operations.

You can find mirror symmetry everywhere you look—leaves, flowers, architecture, ornament. Your own body has, externally, approximate mirror symmetry. Indeed, the bodies of almost all living animals have such mirror symmetry, a fact that can hardly be due to chance. The importance of the concept of mirror symmetry would be hard to overstate. Its role in the mathematical theory of symmetry is fundamental; its role in science is of even broader significance. As one example, Pasteur's discovery of the relation between the rotation of polarized light by certain chemical solutions, and the absence of mirror symmetry in the forms of crystals precipitated from these solutions, was one of the most consequential developments for modern science, paving the way for our present understanding of molecular form. The explanation for his observations is that when both the crystal and the solution rotate polarized light, the individual molecules of the compounds being studied must have no mirror symmetry.

In popular speech, symmetry usually means mirror symmetry, or an approximation of it. However, reflection in a mirror is only one of the many ways of repeating a figure to form a symmetrical pattern. Other ways suggest themselves if we use not one mirror, but two. We can use two mirrors in an ingenious manner suggested by Sir David Brewster, who published in 1819 a small book[2] explaining the history, theory, and construction of an instrument he had recently invented. He called it the kaleidoscope.

A kaleidoscope produces symmetry by optically repeating an arbitrary motif. With two mirrors, which intersect at a properly chosen angle, a beautiful configuration with both rotational and mirror symmetries can be produced. Try it by placing any group of small asymmetric objects, such as bits of broken glass, between two small mirrors. With one mirror alone, we would see the objects and the mirror image of this arrangement, as in figure 1a. But with two mirrors, we see reflection, reflection of the reflection, and so forth, as in figures 1b and 1c. Such a pattern, produced by the mirrors, has not only mirror symmetry but also rotational symmetry. The ever-changing complex of designs that are such a delight to look at in a kaleidoscope are simply the result of a sequence of reflections, in two mirrors, of arrangements and rearrangements of the most ordinary materials.

1. (a) A symmetrical design, produced by reflection in a mirror. (b) Two intersecting mirrors, forming a 60° angle, produce a pattern with rotational symmetry. (c) Mirrors intersecting at a 45° angle generate a different pattern with rotational symmetry.

1. (d) Two nonintersecting mirrors yield a pattern that continues indefinitely. Each mirror is indicated by a straight line. When there is more than one mirror, mirrors are reflected in other mirrors to yield a mirror for each element of reflection symmetry in the pattern.

If two mirrors do not intersect, but instead are placed parallel to each other, a different sort of pattern is obtained. Instead of a circular design, we now have one that repeats indefinitely, like an ornamental border pattern, or a woven band (fig. 1d). With three mirrors, we again have a border pattern; this time the pattern has a mirror plane parallel to the direction of the pattern, as seen in figure 2a. (For the time being, ignore the alternating changing of the shading. We shall return to that feature later.) With four mirrors, we can create a pattern that repeats indefinitely in two independent directions. Thus a plane pattern arises, part of which is shown in figure 1b. A planar pattern can also be created with just three mirrors if the mirrors form the sides of a triangle.

Now you need to use your imagination, because the best drawings can only hint at what happens when even more mirrors are used. With six, forming the sides of a rectangular solid, we can create a pattern that repeats in three perpendicular directions. This is an especially interesting sort of pattern, but it is impossible to view unless one in some manner gets inside the pattern. Because it is three-dimensional, this pattern is not found in ornament. Nevertheless, such unseen symmetries are all around us.

Three-dimensional symmetrical patterns are implicit in many of the designs that are part of our daily lives. Packing arrangements are such patterns. Many modern apartment complexes are built in this way. The atoms in most solid forms of matter are put together in symmetrical three-dimensional patterns. And in crystals the design repeats so many times that one can think of the crystalline state of matter as constructed of infinite arrays of atoms, always arranged in a three-dimensional repeating pattern. The reason why these regular arrangements are used to stack boxes in a warehouse, cans in a supermarket, apartments in a high-rise building, and atoms in a crystal is that in each case the regular arrangement saves space and is stable. The regularity makes planning and construction of the man-made structures orderly, and also permits an orderly approach to the study of the solid state of nature.

By extending the scope of mirror symmetry to repeated reflections in two or more mirrors, we find relations between our own looking-glass

7 Patterns of Symmetry

reflection, kaleidoscopes, ornamental bands, crystals, and apartment buildings. In other types of repetition we can find still other patterns of symmetry.

Rotation

We have already noted that a kaleidoscope pattern has rotational as well as mirror symmetry. This means that the pattern appears the same if rotated through a certain number of degrees about an axis passing through its center; the number of degrees of rotation depends upon the angle between the mirrors. The symmetry *operation* is the motion of rotation, and the symmetry *element* is the hypothetical axis about which rotation occurs. (For a kaleidoscope, this axis is the line of intersection of the mirrors.) If the angle of rotation is 90°, then four consecutive rotations are required to complete the circle of 360°; consequently, this is said to be a four-fold rotation. A rotation through 120° is three-fold, and a rotation through 60° is six-fold.

In a kaleidoscope, the mirrors generate the rotational symmetry: a reflection of a reflection appears as a simple rotation of the original object(s) (fig. 1). But there are also patterns with rotational symmetry that do not have mirror planes. Such patterns can be of several sorts, which we shall point out in plane ornament, in three-dimensional objects, and in motions. The child's pinwheel illustrated in figure 3 is an object with

2. (a) Three mirrors are used to generate a repeating pattern of a special sort; one of the mirrors is "color-active," and reflection in that mirror changes the shading of the pattern. (b) A portion of a pattern is generated by four mirrors, two of which are "color-active." The entire pattern extends indefinitely to fill the plane.

a

b

rotational symmetry (find the rotational axis), but without mirror planes. The symbol shown in figure 4 was designed as the motif for the 1973 Smith College Symmetry Festival. It has a three-fold axis of symmetry, but again there are no mirror planes. Many dances involve rotation motions, often without reflection. The diagrams in figure 5 illustrate some of the patterns of motions for the Pavane, a Renaissance dance. Although we can speak of the symmetries of any instantaneous arrangement of dancers, the symmetries of the motions themselves are the essence of the dance. Thus patterns of symmetry can be produced by rotation alone. In figure 6a we see a pattern produced by rotation about one axis. The pattern in figure 6b results from rotation about two axes. (See if you can find them.)

Translation

In any pattern that repeats indefinitely—whether in one, two, or three dimensions—there is a third type of symmetry element: spatial repetition through a fixed distance. This symmetry is known as *translation*. As we have seen, translation can be generated by parallel mirrors, or by parallel two-fold rotation axes. It can also exist independently, as illustrated in figure 7. Translation in one direction (together with translation in the reverse direction) produces a one-dimensional pattern. Translation in two nonparallel directions characterizes a plane pattern; and in a three-dimensional pattern there are directions of translation which are not coplanar. Parqueted floors, patterned wallpaper, bands of lace, crystal structures, and the yellow brick road all have translational symmetry in the sense that these patterns have no natural boundaries. It is easy to imagine that they could each be extended indefinitely. Translation can be combined with reflection or rotation to produce further symmetry operations. For example, rotation a certain number of degrees, followed im-

 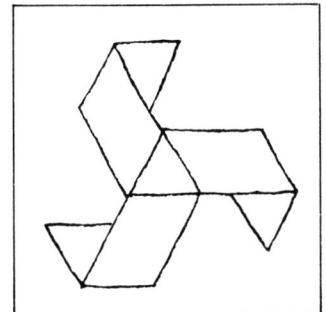

3. (Left) A pinwheel, a three-dimensional object without any reflection symmetry but with a four-fold axis of rotation. 4. (Right) The Symmetry Festival symbol. It may be that the axis of rotation and the absence of mirror planes together suggest, to the viewer, a motion that makes this symbol livelier than a similar figure with reflection symmetry.

9 Patterns of Symmetry

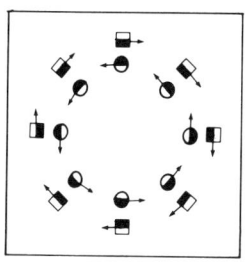
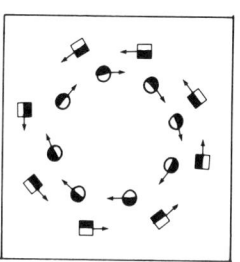

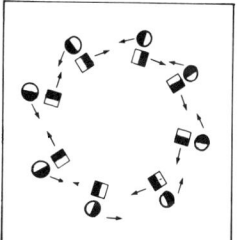
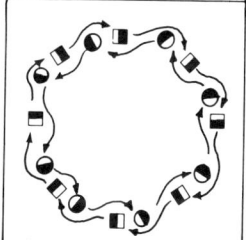
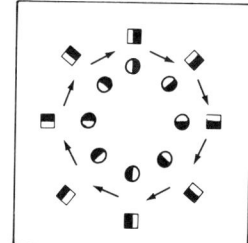

5. A selection of dance patterns from the "Pavane." The symmetrical patterns which the dancers create arise from the rotations of two circles, one of men (▨), one of women (◐). The two circles of dancers rotate simultaneously but independently. The black half of each symbol indicates the front of a dancer. The straight line indicates a mirror, showing that for each pattern there can be imagined another, the two related by such a reflection.

mediately by translation a fixed distance along the axis of rotation, produces helical symmetry, the symmetry of the "spiral" staircase.* Helical symmetry is also found in the arrangement of the leaves of many plants.

Just what combinations of symmetry elements can there be in a single pattern? How many kinds of symmetry elements are there? How many different pattern types exist? We have evidence, in the form of surviving ornament and architecture, that these questions have been studied experimentally in many civilizations. For example, the tombs of the Egyptian kings contain various patterns, the Alhambra in Granada, Spain, is virtually a museum of Moorish symmetrical ornament, and in our own time M. C. Escher ingeniously used old and new patterns in many of his graphic works. These problems have also been considered in various scientific contexts during the past several hundred years. At the beginning of this century it was established that the number of pattern types is in fact

---

* Most so-called spiral staircases are actually helical. In a spiral, the radius of rotation expands continuously, while in a helix it remains constant.

(a) (b)

6. Patterns of symmetry produced only by rotation. (a) The rotation axis passes through the point P perpendicular to the plane of the paper. The angle of rotation is 180° (the axis is a two-fold axis). (b) This border pattern, which extends indefinitely both to the left and to the right, has two axes, one through a point equivalent to P in (a), and the other a different sort of point. Both angles of rotation are 180°.

limited, essentially because the requirement of strict symmetry is a severe constraint.

It is this limitation of the number of pattern types that implies the existence of fundamental relationships among the patterns in all the fields of nature and of the creative arts in which symmetry plays an active role. Thus we find in the ornamental patterns of book design the same elements of symmetry as in tiled floors and in crystal structures. To take another example, we find a relation between ornamental borders and music. In music, the elements of symmetrical construction include the symmetry operations of repetition (translation) and reversal (reflection), and these are the very symmetry elements found in border patterns. While most music is not strictly symmetrical, symmetry operations do form the basis of many compositions. They are especially visible in children's songs, and they probably make the songs easier to learn. They are also found in the music of the Middle Ages and the Renaissance and, often in a very sophisticated form, in many compositions of the Baroque era. Some examples are shown in figure 9. Throughout Western Europe today there are bamboo flute guilds whose members construct their own instruments. It is interesting that the guilds officially urge their members to decorate their flutes with simple symmetrical ornaments, in keeping with both the music played and the form of the flute itself (fig. 8).

7. A border pattern which possesses only translational symmetry.

11 Patterns of Symmetry

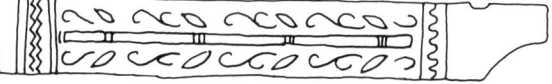

8. Flute decorations chosen to be in keeping with the shape of the instrument and with the nature of the music played. Reprinted by permission from Paul Pfister, "Decoration in Relation with Music," extract from Beatrice Scala, *La Flûte de Bambou* (Switzerland: Guilde Suisse des Flûtes de Bambou, 1965).

12 Introduction

Symmetries of Causes and Effects

Symmetry plays a role—sometimes great and sometimes small—in many fields because of its relation to cause and effect. The problem is often dual: to find the causes that produce a particular effect, and to predict the effects arising from a particular cause. Symmetry as an effect suggests regularity in the origin of a phenomenon or in the formation of a system. Conversely, its absence implies a deviation from regular repetition. Symmetry operations are especially significant as causes because the ideal effects are rigorously predictable, and symmetry becomes a valuable study in itself because of its wide applicability. This duality, cause and effect, leads to two distinct but related starting points for the study of symmetry.

As a first starting point we can begin with the patterns, and search for the processes by which these patterns could have been formed; that is, we can look for symmetry elements. A perfect snowflake can be rotated 60° and no change of position is apparent (we have located a six-fold axis); a symmetrical vase can be reflected in a mirror and we see it reproduced exactly (the vase itself has reflection planes; it is its own mirror image); an endless line of anonymous uniformed soldiers can take one stride forward and the line appears exactly as it did before (we have identified the symmetry element of translation). In every pattern of symmetry we can

9. Selections to illustrate ways in which symmetry can enter into musical composition. (a) An old children's song from Germany showing repetition (translational symmetry) in the construction of a simple tune, with a slight deviation from strict symmetry to add some interest. (b) "Gavotte" by Johann Pachelbel (1653–1706). Here one can imagine a sort of musical helix, with repetition (a repeat-distance of two measures) a third higher in pitch than the original. Again, the composer has chosen to deviate slightly from strict symmetry.

Old German Children's Song

Gavotte   Johann Pachelbel (1653–1706)

13 Patterns of Symmetry

9. (c) "Crab Canon," from *The Musical Offering* by J. S. Bach. This is reflection in a vertical mirror; each violin is playing the other's part backwards. Is strict symmetry observed by Bach? (Reprinted from Bach/Hans T. David, *The Musical Offering*, New York: G. Schirmer, Inc. Used by permission.)

find motions—symmetry operations—with respect to which the pattern remains invariant. In fact, symmetry is often defined as "invariant change." For example, the design seen in the kaleidoscope is invariant with respect to reflection in the original mirrors, to reflections of their reflection, and to the rotational symmetry produced by these reflections.

A symmetry operation is a motion and therefore effects a change in the positions of most of the parts of the pattern; the object which possesses the corresponding symmetry element appears to be unchanged as a result of the operation. One can thus classify patterns by the kinds of symmetry operations that leave them invariant, and in this way discover form. We have learned something important when we note that the most elaborate snowflake has exactly the symmetry of the simple hexagon, no more, no less. We can classify the multitude of images seen through a kaleidoscope whose mirrors intersect at a 45° angle by recognizing that all have the symmetry of a square.

Unexpected relationships can often be discovered by classification, especially when the classification scheme is fundamentally related to structure. Mineral crystals afford striking examples. For example, in the photograph of the mineral fluorite (fig. 10a), we see several intergrown crystals, each of which is a cube or part of a cube. A single, perfectly grown crystal of this substance usually has a perfect cubic form. The regularity of the square faces and the arrangement of the faces to form the solid produce the rotational and mirror symmetries of the cube. These same symmetries are also characteristic of garnet and many other crystals. However, garnet crystals (fig. 10b) appear in nature as rhombic dodecahedra, not cubes. A rhombic dodecahedron is just one of many shapes that have the symmetries of a cube, even though these shapes look very different (fig. 11). Cubic symmetry is characterized by the three, mutually perpendicular, four-fold rotation axes. These axes can be imagined as passing through the centers of opposite faces of the cubes. In the rhombic dodecahedron, the axes intersect the solid where four rhombi meet. The points where three rhombi join together correspond to the vertices of the cube.

As a second approach, we can begin with the symmetry operations and try to determine the symmetrical patterns that will result. You will see this approach in operation in this book, since most of the contributors work with symmetry in a creative way, creating patterns. They ring bells, perform music, design books, create visual patterns. In nature, too, symmetry develops through patterns of growth. Symmetry operations are part of the process of growth, and the symmetrical results are a consequence of

15 Patterns of Symmetry

10. (a) Fluorite. (b) Garnet. Although both minerals have the same symmetry, they have different shapes.

these processes; symmetry arises when similar units are repeated according to a set of rules.

The role that symmetry theory plays in the study of growth and form is most evident in the growth of crystals. Although not alive, crystals do grow. In a solution, in a vapor, or in a melt, they grow from tiny nuclei to visible size. This growth occurs by the accretion of molecular units in a stable symmetrical relation to one another. The shape of the units, and the manner of repetition, is determined by the configuration of the atomic motifs and the directions of the chemical bonds between them. If we know these, we can predict a final external form. If our predictions are not fulfilled, then we look for nonstructural influences which may have disrupted the ideal process.

The two approaches to symmetry are the converse of one another and are abstractly related in the mathematical concept of the *group*. We shall illustrate some properties of a group by performing 90° rotations of a decorative tile (fig. 12).

Let us take a point, P, in the plane of the paper and imagine that a

11. The cube on the left and the rhombic dodecahedron on the right share the same symmetry elements: each polyhedron has three mutually perpendicular four-fold rotation axes, as indicated in the middle.

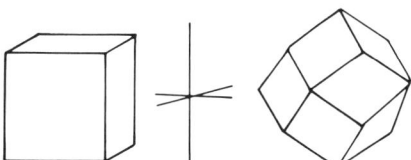

16 Introduction

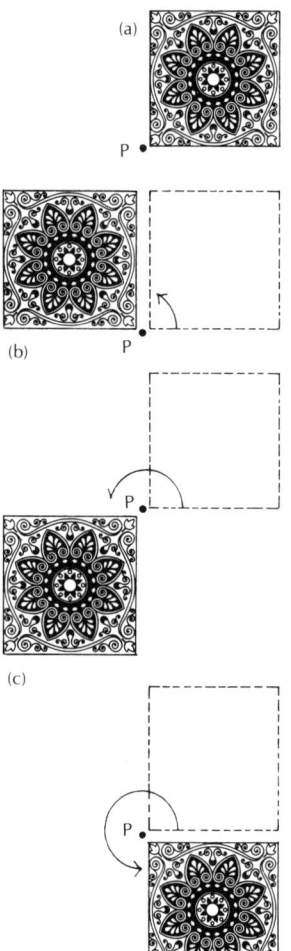

12. Successive 90° rotations of the tile about the point P give four different orientations of the tile.

rotation axis passes through P, perpendicular to the plane. Now let us take a motif—here a tile—and place it near P. The result is shown in figure 12a, and we shall say that any operation which yields this result is equivalent to the operation i, the identity operation. Suppose that the angle of rotation around the axis is 90°, and that the direction of rotation is counterclockwise. We shall denote the symmetry operation "rotation through 90°" by the symbol r. Starting with the tile in its original position, we perform the operation r, and we get figure 12b. If we perform r again, we have in effect rotated the original tile through 180°; the operation of performing r twice in succession will be denoted r^2. Three rotations performed in succession, r^3, bring the original tile into a position below its starting place. And finally, four 90° rotations return the tile to its original position; thus we write $r^4 = i$.

We have created a mathematical system. It consists of an operation r, which we call the generator; and the operations r^2, r^3, and r^4 (with $r^4 = i$), obtained by successive applications of r. The operations r^2, r^3, and r^4 are called "multiples" of r. Rotation through any integral multiple of 90°, counterclockwise or clockwise, brings the tile into a position identical to that effected by one of the four operations: $r^5 = r$, $r^6 = r^2$, and so forth. In this sense the four operations, r, r^2, r^3, and i constitute a complete set of rotations: any product (successive application) of two of them leads to a third in the same set. This system, which has other interesting mathematical properties as well, is a simple example of a group.*

Suppose that we regard r as creating new tiles rather than rotating the original one. Then r generates a pattern: the design in figure 13 is an inevitable consequence of the sequence of operations i, r, r^2, and r^3, and the relation $r^4 = i$. Conversely, we can look at this pattern, observe its four-fold rotational symmetry, and look for a generator of its symmetry group. It is clear that each tile is related to the next by a rotation through 90°. Thus we can locate the generator in a completed pattern.

For every pattern of symmetry, no matter how complex, there is a corresponding symmetry group. The group is usually more complicated than the example above; there will often be more than one generator, but we can always find them all. Conversely, with an appropriate set of generators and a knowledge of the relations between them, we can generate a group and a corresponding pattern of symmetry. The two dual aspects of symmetry are thus united in the mathematical concept that underlies it.

* A precise definition of a group, some of its basic properties, and a discussion of the groups that appear (in disguise) in this book are found in "The Theory of Patterns" in Part 4.

Nonclassical Symmetries

Color symmetry is as old as ornament. Until recently, however, the use of color, tone, or other characteristic was considered extraneous to the mathematical theory of symmetry. The border pattern of figure 1 is an example of color symmetry; as we move from motif to motif, black changes to white, and white changes to black. The discovery that the structures of certain crystals are not completely characterized by the patterns of their atomic configurations, but also must take into account the subpatterns of the magnetic properties of the constituent atoms, led to the new idea that change of characteristic (generalized as *color*) should be incorporated into the general theory along with geometric symmetry. How can this be done?

Let us return to the border pattern in figure 1d for an example. Here we can distinguish between two sets of mirror planes: those passing through the centers of the motifs, and those separating motifs. The mirrors of the first set are "ordinary," because they reproduce both form and color, but the mirrors of the second set are "color-active," reflecting the geometric arrangement faithfully but interchanging black and white. The theory of color symmetry is still in its formative stages. The goal is to determine, mathematically, all possibilities for incorporating any given number of colors in every sort of pattern of symmetry.

The operations of "classical symmetry"—those we have discussed so far—are *rotation* through a fixed angle, *reflection* in an imaginary mirror, *translation* in a specified direction, and combinations of these. We need not, however, be limited to just these operations. The concept of invariant change extends beyond geometric motions, and the processes extend beyond discrete motions to the continuous and even to the impossible.

An example of generalized symmetry is rotation through an angle that

13. Here is a pattern which can be considered to have been formed by the symmetry operations illustrated in Figure 12. It has a four-fold axis of rotation.

does *not* evenly divide 360°. Combined with translation, this creates a form of helical symmetry that cannot be a symmetry element of a crystal (the reasons are discussed in Mrs. Senechal's essay in Part 3) but has been found to be a key feature of many complex, biologically important molecules, including DNA. Continuous rotation and continuous expansion of the radius of rotation together sweep out a spiral; this is the symmetry we find approximated in shells, in the growth patterns of many crystals, and, combined with translation, in the shapes of the horns of many animals.

As another example, we can take unperformable symmetry operations. Let's look at a cube. If the vertices of a cube are numbered, as in figure 14a, the symmetry operations of the cube can be described as permutations of these numbers. For example, a four-fold rotation about an axis through a pair of opposite faces changes the positions of the vertices: where 12345678 used to be, we now find 23416785 (fig. 14b). But there are many rearrangements, such as 21435867, which cannot be obtained by any symmetry operation of a normal cube (fig. 14c). In a general sense though, every rearrangement of the eight digits is an instance of invariant change: the positions of the numbers are changed, but the set of numbers itself is invariant, since all eight digits are present and no new ones have been added. The symmetry operations of the cube can be regarded as a sub-group of the group of all permutations of the digits 1 through 8.

"Impossible" symmetries appear in the patterns of Renaissance (and other) dances and in change ringing, because dancers can move and bells can be rung in ways which have no analogues in the symmetry operations of rigid polyhedra. These generalized symmetry operations might be contrasted to the "real" ones, but this distinction is not so clear as might first appear. Although all symmetry operations are regarded as "motions," the only operation of classical symmetry that is actually possible, in the sense that it can be performed on a three-dimensional object, is rotation. The familiar mirror symmetry cannot really be carried out: there is no way to change a right-handed glove into a left, although we describe such an operation using the analogy of a mirror. Translation is also impossible (one can move an object a fixed distance, but then its position is changed)

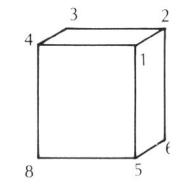
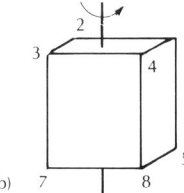
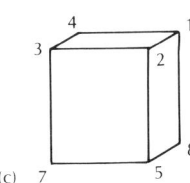

14. Possible and impossible symmetries. Note that the cube in (a) and the cube in (b) are related by a 90° rotation about the indicated axis. But there is no symmetry operation that could convert either into the cube in (c).

unless the pattern is infinite in extent. Thus there is no fast line to be drawn between the possible and the impossible, the classical and the new, and we will not try to make such sharp distinctions.

Finally, there is a more subjective generalization of symmetry: a liberal attitude toward strictness of interpretation. We can find similar patterns of symmetry in entirely different fields of endeavor. It is used in different ways, and in different degrees, for many different purposes. In some cases we recognize the patterns easily, while in others careful study or a vivid imagination is required to find them. In all cases, recognition of the similarities between the many patterns makes possible a deeper appreciation of their differences.

Seek and Ye Shall Find

In Bach's time, when symmetry was an important principle in composition, musical puzzles were a popular intellectual game. One of these was the solution of "puzzle canons." The composer provided a theme and the task was to determine by what symmetry operations he intended the theme to be repeated. There are three of these puzzle canons in the *Musical Offering*. The challenge was posed by the inscription Bach wrote on one of them: *Quarendo Invenietis,* Seek and Ye Shall Find. Nature presents us with puzzles of the opposite sort—it is as if the completed canon were given and we are to discover the theme and the pattern—but the inscription applies equally well. Thus we find again the twin faces of the study of symmetry: *seeking* the rules and motifs through which existing patterns of symmetry have arisen, and *finding* the patterns which arise when a motif is repeated according to various rules. The first leads to the study of the structure of matter, art, music, and thought. The second poses questions of design which have intrigued artists, architects, musicians, and scientists since ancient times. These two facets of symmetry are explicit or implicit throughout this book. The motifs include leaves, the sounds of bells, threads, molecules, typographic arabesques, colors, and dance steps; the patterns are called foliage, change ringing, weaving, crystals, typographic ornament, color symmetry, and branles.

Seek and Ye Shall Find. Our purpose is to demonstrate the significance and excitement of the search. No puzzles in music, art, or science are solved in this volume. Rather, we exhibit facets of nature and human creativity in many fields in order to illuminate common patterns of symmetry that might otherwise remain hidden. We invite you to explore these patterns. The variety of motifs may at first be bewildering, but the apparent chaos will soon resolve itself into order, as the same patterns of symmetry are repeated over and over again.

The Many Facets of Symmetry

It was early February 1973, and the glass-enclosed connector between the buildings of Smith College's Clark Science Center had been transformed into a walk-through introduction to symmetry. In every other window, on each side of the connector, were hung identical green and yellow symmetrical symbols; a green line down the middle of the floor emphasized the mirror-image relation between right and left walls. Those looking at the connector from the outside could also see that the repetition of the designs on the two floors created a more complex symmetry known as glide: each floor was related to the other by mirror reflection and a parallel shift.

The highly visible connector was introducing the 1973 Smith College Symmetry Festival, a Vanderbilt Symposium held February 16 to 20. The symbol used in the connector appeared on posters and banners, heralding the coming of five days of lectures, dancing, music, discussion, bell ringing, and exhibitions. The Symmetry Festival was planned and directed by the editors of this book and a colleague, Allan Ludman. The idea originated in our exploration, over a period of years, of the many concepts common to our own disciplines (mathematics, chemistry, and geology), and in our sharing of ideas and resources. Through these events and exhibits, we hoped to demonstrate a principle of organization which underlies many areas of science and the humanities: the repetition of small parts in prescribed ways to create a whole which is more than their sum. In a word, symmetry.

The major exhibition of the festival, "The Many Facets of Symmetry," was displayed in McConnell Foyer, Clark Science Center. There one could see how an alum crystal grows as molecules in solution arrange themselves in patterned layers to build up the regular plane faces of an octahedron, and how a textile design grows on a loom as the harnesses are lifted according to a plan and the weft is laid down row after row. The intricate symmetries of architecture appeared side by side with no less intricately symmetrical plants from the Lyman Plant House. The visitor

This chapter is based on the following: the essay "Symmetry Festival," by Marjorie Senechal, which originally appeared in the *Smith Alumnae Quarterly*, April 1973, pp. 14–16; an essay on dance prepared for *Patterns of Symmetry* by Arthur L. Loeb; and the essay "Symmetry in Music," written for the Symmetry Festival Guidebook by Lisa Compton, Smith College Class of 1973.

1. (Left) Walking through an environment with reflection symmetry. The intrinsic symmetry of the Clark Science Center connector is emphasized by Symmetry Festival symbols hung on each side, and a line that locates the symmetry element for the corridor and its decorations.
2. (Right) Two school children, visiting a Symmetry Festival exhibit, constructing models of molecules in an attempt to make asymmetric nonsuperposable mirror-image models of molecular isomers. (Photographs by Gordon Daniels, reprinted from the *Smith Alumnae Quarterly*, April 1973, and used by permission.)

was drawn to tables at which she or he could sit down and study symmetries, patterns, and polyhedra, and create her or his own designs. In addition, there were mobiles and models of many shapes and sizes, an electronic sculpture of light reflected in mirrors, and a series of panels showing the repetition of simple motifs according to the requirements of color symmetry. Other exhibits of the festival appeared in other parts of the Science Center, in the Museum of Art, and in the Neilson Library.

Professor Arthur Loeb, a scientist, musician, and designer, gave the first lecture of the festival on Friday, February 16. This lecture, "Color Symmetry and Its Significance for Science," was a joint open meeting of Mathematics 110b *(Introduction to Symmetry)*, Chemistry 241b *(The Structure of Molecules)*, and Geology 221b (the second half of *Mineralogy and Petrology*). He showed how symmetry operations can be combined, and indicated some of the combinations that produce further symmetries. He then discussed how his approach to the exploration of pattern construction can be extended to create a self-consistent theory of color symmetry. The text of his remarks appears on pp. 28–43.

The second event was a lecture-demonstration by the Cambridge Court Dancers, who presented a program of Renaissance Court Dances under the direction of Ingrid Brainard. Ingrid Brainard discussed the history of

The Many Facets of Symmetry

the dances, and Arthur Loeb, one of the performers, explained how the highly structured units of the dances are combined to produce visual designs of complex symmetry.

In these dances, relationships between the dancers represent interpersonal relations, leading, in the *balli*, to the telling of a short story. For example, in a couple dance, "La Malgratiosa,"[1] the courtier describes a rectangular path around the ungracious lady. The first two sides of the rectangle represent the courtly attempts of the man to impress his lady, while the other two sides represent his increasing fury. The steps are different in these two parts of this dance. Finally, the courtier crosses diagonally toward the center of the dance to carry off the unwilling coquette.

In "Verceppe,"[2] a ballo for five dancers, the symmetry is entirely two-fold rotational. At the center of symmetry we find Scaramuccia; in front and in back of him are two other couples of the Commedia dell'Arte. The entire group of five moves in a line, forward and backward. The two men and women change places from time to time, always preserving the two-fold symmetry, dancing around the central Scaramuccia.

The *basses danses* and *basse danze* are primarily processional dances. All dancers are in couples, lined up one behind the other. The only interactions are between partners; the symmetry is translational. More complex is "La Ligiadra,"[3] a *ballo* for two couples. Here we see a flirtation between dancers of the opposite sex and different couples. Every dancer relates to his own partner, to his rival, and to his rival's partner. After an initial procession, the leading couple turns around in place, woman facing woman, man facing man. The structure has color mirror symmetry:

3. Undergraduate students investigating mirror symmetry, construct symmetrical patterns by manipulating a pair of mirrors. They are using the principle of the kaleidoscope. (Photograph by Gordon Daniels, reprinted from the *Smith Alumnae Quarterly*, April 1973, and used by permission.)

first man	first woman
second man	second woman

mirror

color mirror

At the intersection of a color-active and a color-inactive mirror is a color-active, two-fold center of rotational symmetry. The first man and second woman meet at this point for a flirtation, then change places along an S-shaped curve. The second man and first woman follow suit. There is strict adherence to two-fold symmetry. The S-shaped curve causes the rivals to turn their backs on each other, but eventually the partners turn in toward each other, and conclude the dance in a gesture of reconciliation.

"La Gelosia"[4] is a dance of jealousy. Three couples enter, one behind the other. The first man then leaves his partner, flirts with the second woman, and winds up with the third. Meanwhile second and third men move up, respectively, to first and second women. In successive rounds of dance, the men permute cyclically between the women, who remain in the original sequence.

All of these dances date from the fifteenth century, The *Pavanes* and *Tourdions* are processional dances from the sixteenth century that have translational symmetry. The *branles* also have symmetrical structures.

Each of these dances is characterized by its symmetry. Very rarely does one find configurational changes involving an alteration of symmetry in the middle of the dance. The configuration of the dances, and the numbers of dancers involved, all varied greatly in the Renaissance. Each dance, however, has a specific symmetry, which is a dominant structural determinant. The performance was one of the most popular of the festival, attracting more than 500 people who crowded into Scott Gymnasium Saturday night. Over a hundred stayed on afterward to take lessons in Renaissance dancing from members of the group.

On Monday afternoon a wide variety of people gathered in the Browsing Room of the Neilson Library to discuss problems of interdisciplinary work with Arthur Loeb, under the title, "A Renaissance Man in a Competitive Culture?" These same questions were raised again at a second discussion at Smith College on May 1, 1974; some excerpts from the 1974 discussion appear on pp. 144–46.

Symmetry is a vital part of any aspect of music, from the arrangement of monumental symphonies to the minute structure of melodic phrases. It creates order, providing a pattern which unifies whatever diverse elements the composer chooses to use in his work. Because of its intricately symmetrical structure, *The Musical Offering* of J. S. Bach (BMV 1079) was chosen for performance on Monday evening by faculty and students of the Smith College Department of Music.

A symmetrical placement of contrasting and repeating sections in a composition was typical during Bach's time and during the Classical period. For example, the ternary form (ABA) appears in such genre as the *da capo* aria, some dance forms, and modified in the sonata allegro. The first A section states the theme or themes. The B section either develops it or contrasts completely; the final A section recapitulates or repeats the beginning theme(s). Many other symmetrical forms exist, for example the rondo (ABACABA). Contrasts of elements other than themes are important also, such as tempo, tonality, and dynamics. Beginning a piece in one key, developing it through others, and returning to the starting one was widespread during the Classic Era.

The Musical Offering is made up of thirteen parts. These are arranged in five-part symmetry, each with an internal symmetry of its own:

   I   Ricercar
   II Five Canons
   III Trio Sonata
   IV Five Canons
   V   Ricercar

Although made up of short sections, this composition is unified in several ways. All these sections are based on the same theme, either stating it literally or alluding to it. Bach developed and changed its shape in countless ways; he was a master of arranging proportions of contrasting sections into a balanced, often strictly symmetrical whole. The symmetry of this work is evident not only in its outer structure but also in the units of the pieces themselves. For example, the various canons are the musical analogues of ornamental patterns of floors, walls, and textiles, in which the second row is symmetrically related to the first. The ricercar, in common usage by the Flemish and Venetian schools around 1600, was a form involving improvisational treatment of one theme in imitative voices. It soon developed into the fugue as Bach knew it. The word ricercar originally meant "to search." *The Musical Offering* was performed by Philipp Naegele, violin; Marcia Weinfeld, violin; Ernst Wallfisch, viola; Adrian Lo, viola; Diane Berthelsdorf, violoncello; Peggy McGirr, violoncello; Helga Kessler, flute; and Lory Wallfisch, harpsichord.

The festival closed on Tuesday with an afternoon discussion and demonstration of change ringing by Alice Dickinson of the Department of Mathematics and four student ringers, and an evening lecture by Elliot Offner of the Department of Art on "Renaissance Typographic Ornament: Origins, Uses, and Experiment." Alice Dickinson described how the desire of the English to ring their bells in an orderly way led to a method that has strong analogies to the motions of some of the Renaissance dances, and to compositions which involve symmetries in each part and in the whole. The construction of designs from the repetition of a single motif was the theme of Elliot Offner's talk, which he illustrated with slides showing the Granjon and other arabesques (typographic motifs) and some ornaments that had been created with them. Both lectures appear in this section.

ARTHUR LOEB

Color Symmetry and Its Significance for Science

The Symmetry Festival grew, just as the crystal is growing in the exhibit in McConnell Hall. It grew from a little seed some years ago, and it is a tremendous credit to the three organizers—George Fleck, Allan Ludman, and Marjorie Senechal—that it is really off the ground. I know that it was a great effort. They deserve congratulations.

This is the world's first symmetry festival. Long ago, some say, there were plans for an earlier one, but it never took place. Figure 1 shows what happened to those plans. We see here the attempts to design a mosaic which was to be the motif for the festival. It was supposed to be a floor mosaic design in which the whole floor would be covered with regular hexagons and with six-cornered stars. In technical language, every hexagon and every six-cornered star (hexagram) had to be located on a six-fold center of rotational symmetry. That was the first assignment for the world's first symmetry festival. It couldn't be done because, although every hexagram is surrounded by six regular hexagons (this is what we mean by a six-fold center of rotational symmetry), every hexagon is surrounded by only three stars and by three other hexagons. There's no way around it; they kept trying and trying, but the festival never got its motif because it is not possible to meet both of these requirements. Before the end of this discussion we will know why.

For another example of a repeating pattern, look at figure 2. We have here a sort of stylized flower with eight petals, but it is not possible to have eight-fold symmetry in the pattern. As you can see, everything around the center here repeats *four* times, and if we are going to have a periodic pattern extending to infinity we can, at most, have four-fold repetition about this motif. Although the motif, in isolation, would have eight-fold symmetry, its environment allows it only four. The centers of the spirals are also four-fold, because there is a repetition of four—frequency four, you might say—around them. Then we have two-fold centers at the midpoints of the imaginary lines between identical four-fold centers. We have four, four, and two. Note that their reciprocals, $\frac{1}{4}$, $\frac{1}{4}$, and $\frac{1}{2}$, add up to one. Is this significant? We don't know yet, but we'll find out.

In figure 3, one finds a center of six-fold symmetry, a center of three-fold symmetry, and a center of two-fold symmetry at the intersection of the line joining two three-fold centers and the line joining two six-fold centers. And if we take six, two, three, and add their reciprocals, $\frac{1}{6} + \frac{1}{2} + \frac{1}{3}$, we again get unity.

Figure 4 shows one of M. C. Escher's creations. Six lizards are grouped together about the point where their paws touch, and this pattern is also six, two, three. We must look carefully, because here is something that we haven't encountered before: we have white lizards, red lizards, and black lizards. This is very significant because it introduces a concept that was not taken into account in the nineteenth century when symmetry was first studied extensively. The nineteenth-century symmetry theory dealt with identical objects that had to be identical in all respects except their position. Here we have lizards with exactly the same shape but with different physical attributes.

Professor Caroline MacGillavry, a crystallographer at the University of Amsterdam who wrote a book[1] on the symmetry aspects of M. C. Escher's work, believes that Escher was the inventor of color symmetry. Escher, with his characteristic modesty, might laugh and say, "Isn't that silly. They think I've invented it, but how could I do anything else? Obviously, if I made everything the same color, you wouldn't be able to have a boundary to any of these objects; it would be all white, or all black, or all red." But no one before Escher thought of making his background equivalent to his foreground; that is the very novel thing about his work. He fills the plane with motifs without any space in between to separate them. Indeed,

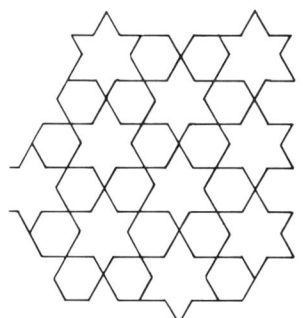

1. A mosaic design which could cover a floor or a whole plane. Every star is surrounded by six hexagons; every hexagon is surrounded by three stars and three other hexagons.

2. Stylized flowers, each with eight-fold rotational symmetry, used as motifs for a plane pattern. A plane cannot be covered by an eight-fold symmetry pattern; the motif and its environment together (that is, the whole pattern) have four-fold symmetry. Reproduced by permission from P. Fortova-Samalova, *Egyptian Ornament* (London: Allan Wingate, 1963).

3. In this pattern are six-fold axes, three-fold axes, and two-fold axes. Where are they?

if it's your major premise that you are going to make your foreground and your background equivalent, then you would be bound to invent color symmetry. Making the background and the foreground equivalent is Escher's major contribution.

I will try, very quickly, to give an outline of my approach to the symmetry problem. It is a flatland approach; that is, I've only looked at the plane so far. My book[2] is concerned with the plane; the idea was to exhaustively generate all the possible configurations of symmetry and color symmetry elements in two dimensions, and to prove that it was exhaustive. (I have to add that in color symmetry we're in a very new territory. There hasn't been any kind of consensus of definition, so I had to work out my own.) I had, much earlier, studied the classical theory of plane symmetry with Professor Le Corbeiller at Harvard. We were a little puzzled about the sort of *ad hoc* nature of some of the theory for the ordinary symmetry in the plane, and we had come up with something like a proof of exhaustiveness which we rather liked. We presented the proof at Rome at an International Congress of Crystallography, and on the way back I visited Professor MacGillavry in Amsterdam while she was correcting the proofs of her book on Escher's graphics. She remarked that Escher had some patterns that had not previously been reported by the Russians (who had already done so much with color symmetry). "It's remarkable that Escher has some more," she said. So I said, "Well, there must be some way of tracking down why they missed these, and what's new about Escher's."

31 Color Symmetry and Its Significance for Science

4. *Lizards*, by M. C. Escher. The symmetry depends on whether one considers the color of the lizards. Reproduced by permission of the Escher Foundation, Haags Gemeentemuseum, The Hague.

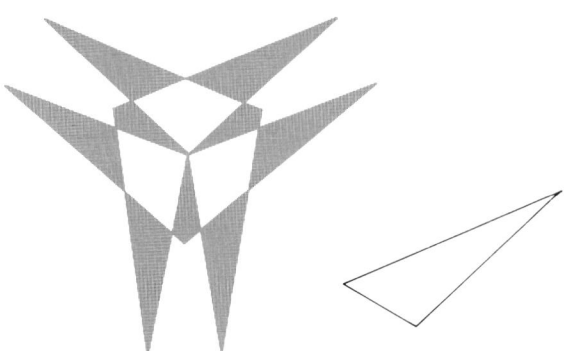

5. Creating a pattern with an asymmetrical motif. The pattern on the left is produced by performing a reflection and two rotations on the white motif on the right. Overlapping regions are white.

My approach to the whole area, both to the classical problem and to the color symmetry problem, is very much conditioned by the fact that I do quite a bit of work in the computer field. I use an algorismatic approach. An algorism is a generating rule. In particular, rather than analyze patterns and hope that I get them all, I start with two symmetry elements and find out what happens when they interact. In figure 5 on the right is a totally asymmetric figure that I had interact with a three-fold roto-center (that is, a center of rotational symmetry) and a mirror. There is a vertical mirror in which the original motif is reflected, and then the resulting pair is rotated to get another pair, and finally a third pair. Then I colored the sections that overlapped when rotated. This design was the emblem for a conference we held some years ago.

Figure 6 shows another algorism. I started with a motif (a triangle), and I had that interact with a glide. In interacting with a vertical glide line,* every triangle is reflected again and again. I deliberately made them overlap so that we get something a little more dynamic looking. Now in figure 7 we have a motif which itself has two-fold symmetry. I had that motif interact with a mirror; you see that that gives the mirror image of the original. Then the two-fold center interacts with this mirror. So just two elements—a two-fold center and a mirror—give us an infinite pattern. It goes infinitely up and down; it doesn't go to the left or right. Again I used overlap and I thought that it would be visually interesting to shade the in-between areas; the result is that the two-fold center and its mirror

---

* A glide line is the symmetry element corresponding to the compound symmetry operation of reflection followed by translation. The mirror is parallel to the direction of translation.

image are emphasized. I find, artistically, that I like the shape of this because the outline is still there, giving it a certain balance. If I had just used Z-shaped figures—or any shape at all—I don't think I would have liked the pattern quite so well. This way, you *see* the interaction: a two-fold center, its mirror image, the original center, its mirror image; the mirror is there, and a perpendicular glide line is implied. The glide line is a new element generated by the first two.

Now let us consider an assignment: at the right in figure 8 there is a rectangle with two-fold symmetry as well as mirrors, and a triangle with three-fold symmetry as well as mirrors. I am to juxtapose them, and my requirement is that I make a pattern in which every rectangle maintains its

6. (Left) A repeating pattern generated by interacting a triangle with a glide operation with simultaneous reflection and translation. Shaded regions of overlap. 7. (Right) Using a parallelogram as a motif, a repeating pattern has been generated by interaction with a two-fold rotation axis and a mirror. The original motif and the infinitely extending array of axes and mirrors are shown on the right. The resulting pattern, on the left, has been shaded.

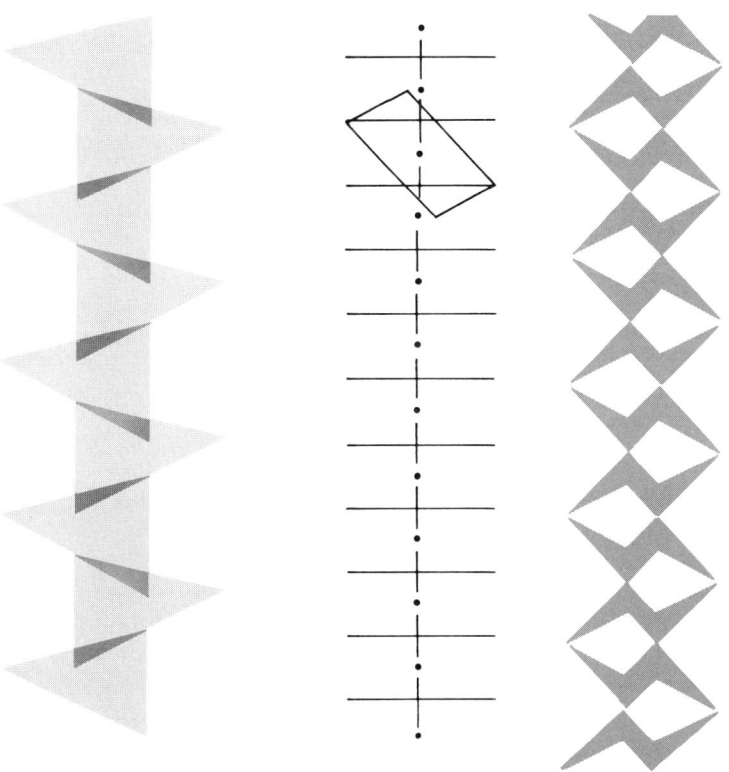

8. A rectangle and a triangle (at the left) are used to create patterns in which every rectangle maintains its two-fold rotational symmetry, and every triangle maintains its three-fold symmetry.

two-fold symmetry and every triangle its three-fold. That means that every triangle has to be surrounded by three rectangles and every rectangle by two triangles. (I've lost the mirrors; that has to happen in this sort of juxtaposition.) Now in figure 9 we find out what happens as we continue. Notice that there is a certain similarity to the Symmetry Festival emblem (p. 14), but there is also a difference. In the emblem there is a similar juxtaposition, but triangles adjacent to a parallelogram are not related by two-fold symmetry. In figure 8 I have them juxtaposed in a way that preserves rotational symmetry. The fact that this is a rectangle and not a parallelogram is totally immaterial here; the important point is that the juxtaposition is fundamentally different from that of the festival emblem. With my algorism, I generate the pattern in figure 9. We begin with two-fold and three-fold rotational symmetry, and what happens? In the background we find a six-fold rotocenter. (Again we have two, three, and six; and $\frac{1}{2} + \frac{1}{3} + \frac{1}{6} = 1$.) That is what I mean by an algorismatic approach; I start with only two elements and I generate an infinite pattern with all its implied symmetry elements. What I want to do, then, is to derive systematically all the possible pairs or triplets of interactions. Then I should be able to exhaustively generate all possible patterns.

In figure 10 we have a square—four-fold symmetry—and I want a second square to be juxtaposed so that they are related by two-fold

35 Color Symmetry and Its Significance for Science

symmetry. To preserve its original four-fold symmetry, each square has to be surrounded by four others; each one is surrounded by others in exactly the same way. (For color symmetry, I would use, say, blue and red squares). The second four-fold center is implied here, and again we find: $\frac{1}{4} + \frac{1}{2} + \frac{1}{4} = 1$.

In figure 11 we again have a three-fold center and its mirror image, also three-fold; a third three-fold center is implied. This one is at the center of a hexagon. It's not a regular hexagon; it has three right angles and three very large angles, so its symmetry is three-fold instead of six-fold. This pattern is three, three, three; and $\frac{1}{3} + \frac{1}{3} + \frac{1}{3} = 1$. The first one and its mirror image are distinct; they turn in opposite directions.

Now, how can I make sure that I get all the combinations? How can I make sure that I know in advance what combinations of rotational symmetry will or will not work? Mrs. Senechal discusses viruses in her essay on order (p. 91) and makes the point that five-fold symmetry can neither replicate itself nor have parallel axes. How do we know that five is not going to work? Well, let us take a *k*-fold rotocenter and an *l*-fold center,

9. (Left) The three patterns shown are continued here in a manner which would, if extended, cover the plane. The repeating motif is indicated at the left. The pattern can also be generated by repetition of the parallelogram-shaped design unit outlined at the lower right. 10. (Right) A motif, constructed from two squares to have only a two-fold rotational axis, generates a repeating pattern with the full four-fold symmetry of the square.

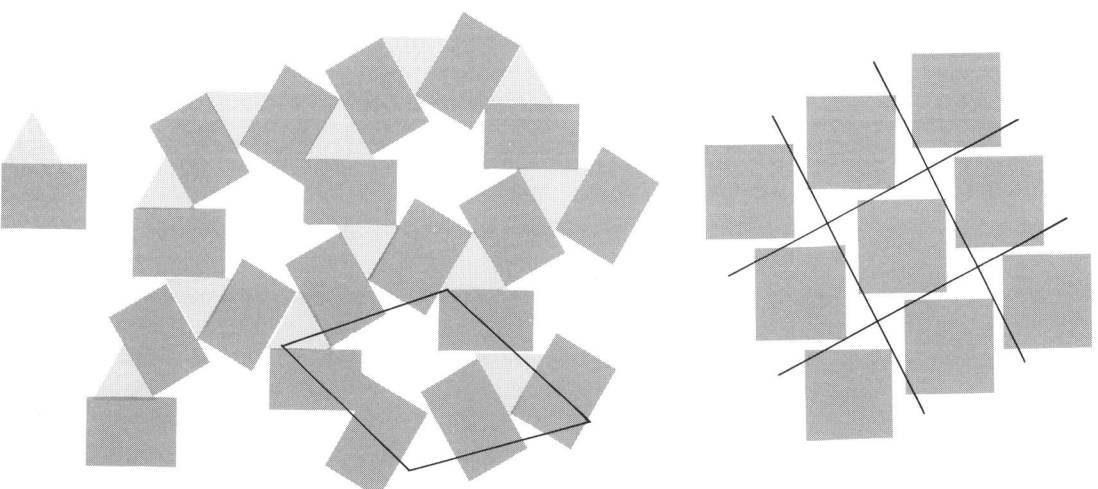

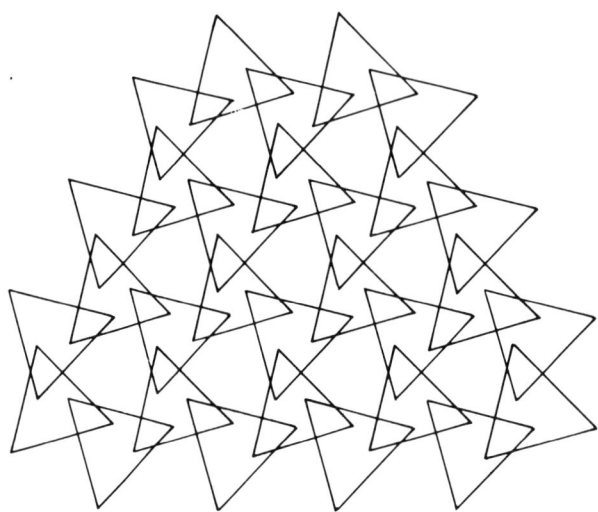

11. This pattern is not the same as its mirror image. Can you find the hexagon in the pattern? This hexagon has only a three-fold rotational axis (not a six-fold axis). Why?

where k and l are unspecified whole numbers. Each of those requires a repetition around itself in the entire plane; that means that around an l-fold center I must have l k-fold centers, and around every k-fold center I must have k l-fold centers. In figure 12, there are three k-fold centers and two l-fold centers between them. Notice that the centers lie on two concentric circles. The process shown here can be continued indefinitely. When you come around full circle, will you match up at the starting point, or will you overshoot? Under what conditions will you just match exactly? Let me answer this in an oblique way. I would say, "Well, suppose that we just overshot, so that we would have a new l-fold center in between two which are already on the circle; I could have it interact with the nearest k-fold one and start the whole story again on a smaller scale, going around and overshooting again, and repeating the whole thing again and again until eventually the centers would come arbitrarily close together. Now in mathematics this is not unheard of: you would have an example of a so-called continuous group. But in crystallography, in ornamental art, and so forth, we want discrete patterns with discrete motifs. We couldn't possibly allow them to come arbitrarily close together. And for this reason I introduce a very explicit postulate, namely, there must be a minimal distance of approach between rotocenters. And once we postulate that, we see that the cycle must close on itself to give us a symmetrical semi-regular polygon (not regular, because alternating vertices have different angles). That means that in the center there must also be a rotocenter—that's the implied one we saw in the different patterns, in the background. Now in figure 12 the angle at A_k is $2\pi/k$, and the angle at A_l is $2\pi/l$; the one in the center, at C, must be $2\pi/m$, where m is the implied symmetry there. When we add all these angles together, we know that their sum must be 2π, or 360°: there are two triangles each of whose angles add up to 180°. So when we add them together and divide by 2π we get

a) $$\frac{1}{k} + \frac{1}{l} + \frac{1}{m} = 1$$

Thus the coexistence of k-fold, l-fold, and m-fold centers of rotation is constrained by the equation (a), whose five solutions can be designated by letters, as follows:

U: 1, ∞, ∞
D: 2, 2, ∞
Q: 2, 4, 4
T: 3, 3, 3
S: 2, 3, 6

Since there are only five solutions, we can exhaustively plot our patterns, as I'm going to show very quickly. The U solution has no real rotocenters, only translation, so we will skip that one. Here are the nets for the others.

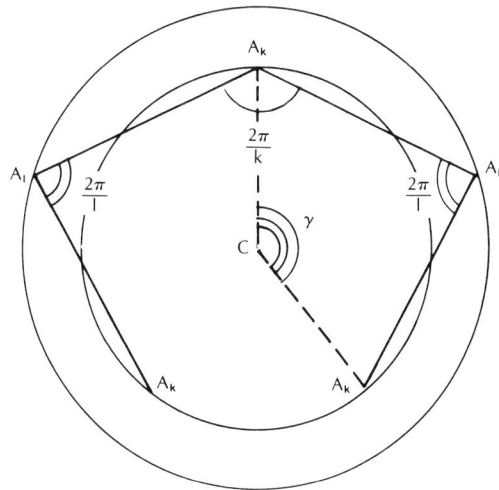

12. Diagram to illustrate the results of various values of the numbers l and k.

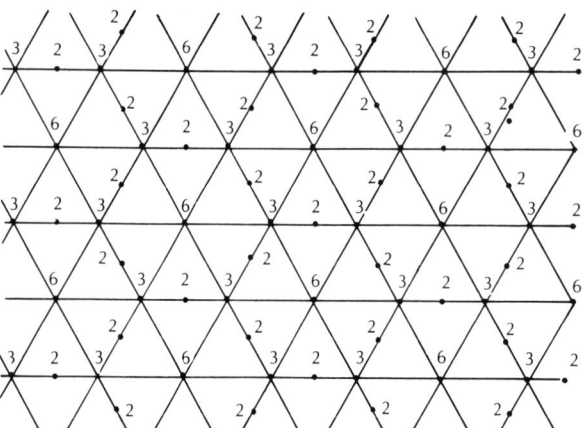

13. The S net.

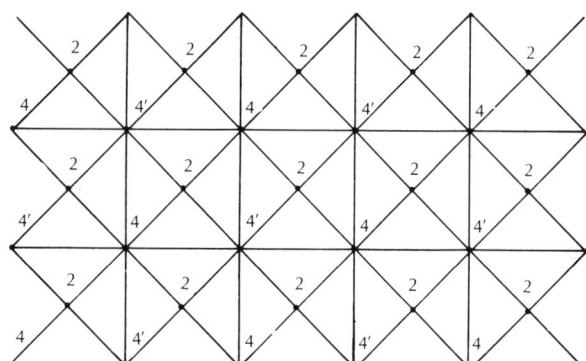

15. The T net.

14. The Q net.

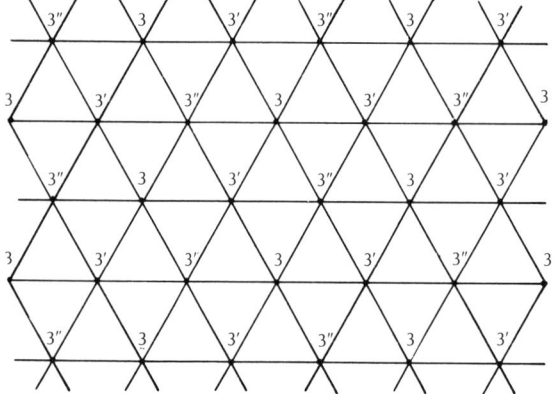

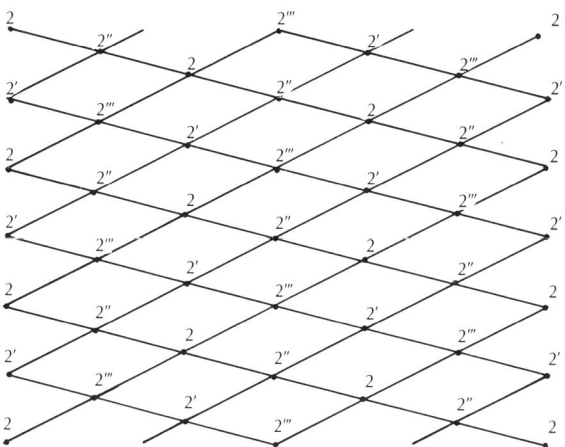

16. The D net.

You can see the completed polygons. Figure 13 is the 2, 3, 6 solution. Figure 14 shows the 2, 4, 4 solution, and figure 15 shows the 3, 3, 3 solution. Figure 16 shows the 2, 2, ∞ solution. These three rotocenters (counting translation as a rotocenter) lie on a straight line, and we can put a lot of them parallel to each other. As a consequence, the net has four distinct rotocenters. With this information one can place motifs on all these nets.

These are the plane patterns of classical symmetry theory. How do we go from here to color symmetry? We know that a rotocenter requires a repeat of equivalent points, of motifs if you like. If you have a point somewhere in the vicinity of a five-fold rotocenter, say, then we know that that point occurs in equivalent position five times. When we go to color symmetry and make that rotocenter "color-active," we assign a hierarchy of colors to each of these points. They could be all different, or some might be different and some might be the same. But there has to be some sort of order, and that's why I had to make some arbitrary definitions.

The most important thing to remember is this: if two points of the same color are acted upon by a color rotocenter or by a color-active mirror so that their color is changed, then their new colors should also be identical. If I have two red points and they are acted on by a color-active center, they might both become blue, or both green or both yellow, but not one blue and one yellow. That's crucial; otherwise there is chaos. As a corollary, if the two points had different colors to begin with, then they shouldn't have the same color afterwards. That is my *consistency postu-*

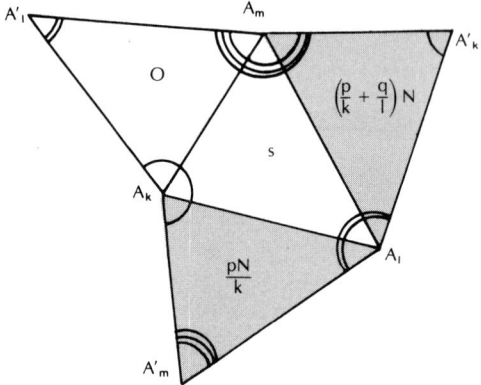

17. Four adjacent meshes in a net.

late. It tells me that as I go around the circle, around the rotocenter, I must have an orderly hierarchy of colors. Since they are ordered, I can assign them numbers; for examples, 0,3,6,9. You might ask, why don't I just call them 0,1,2,3? The reason is that there might be other colors in the pattern not directly concerned with that rotocenter, but with another. In general what I have to do is to go in steps of $0, s, 2s, 3s, 4s,$ in such a way that I end up with 0 again. That means that I must define a number N in the pattern, and then I have to do all addition modulo N.* And when I do that, I find (fig. 17) that I go up by steps of pN/k, where k is the rotational symmetry value, N is the total number of colors in the pattern, and p is an integer. It's a true quantum number; those of you who have studied some wave mechanics remember that your wave function has to be identically itself again when you go all the way around a circle; similarly, color has to be identically the same.

Now when I consider color rotocenter interactions, I go through exactly the same kind of process as I did before: I go around the k-fold center; I proceed down my list of colors by the amount pN/k. Then I go around the l-fold center and proceed by the amount qN/l (q is the "quantum number" associated with the l-fold center). Then by going around the m-fold center I come back to where I started, so the color must be the same as the original color. So we have $p/k + q/l + r/m$, all times N, and that must be 0, modulo N. This leads us to a second equation,

b) $\quad \dfrac{p}{k} + \dfrac{q}{l} + \dfrac{r}{m} = 1$

* That is, all multiples of N are subtracted from the sum. Addition modulo 12 is used in telling time: four hours past 11:00 is 3:00, not 15:00.

Symmetry values	Rotational color parameters
1∞∞	—
22∞	000
	00 (½ m)
	111
236	000
	014
	103
	111
244	000
	013
	022
	102
	111
333	000
	012
	111
2222	0000
	0011
	1111

18. Symmetry values and color parameters in the plane.

Now you see that we have equation (a) for the geometric symmetry, and equation (b) for the color "quantum numbers" of the rotocenters. Together, these are the basic equations for color symmetry in the plane; if I solve them simultaneously then I have, in principle, all the possible combinations (fig. 18). The conclusion is that from these equations I can then exhaustively generate all the patterns.

Now the question is, what is the significance of color symmetry for science? Well, in scientific applications we go beyond color as color; we use color as a code to denote other physical properties. For instance, some crystals have a magnetic dipole moment, magnetic materials in which little magnets are located on atoms—iron atoms let us say. X-rays were first used to analyze crystals, and X-rays are what you might call color-blind to magnetic dipoles. They are insensitive to magnetism; they are photons—light—and they will not interact. So an analysis by X-rays of magnetic crystals would simply show us the location of the iron atoms but nothing about whether they have dipoles and, if they do have dipoles, how the dipoles are oriented. Whereas, if one used a neutron beam (neutrons are particles which themselves carry a magnetic moment, a little magnet), then those magnets on the neutrons will interact with the magnets of the atoms, and we can therefore tell which atoms contain magnets and in which direction these magnets are oriented. It was in the study of magnetic materials, comparing neutron diffraction patterns with X-ray diffraction patterns, that scientists began to realize that there must be some property, in addition to just position, that determined the diffraction of the neutron beams. So they said, "Well, let's denote this by colors."

Another example is the twinning of crystals. Here you have planes which divide one perfectly periodic structure from another identical structure; they come together at a sort of mismatch. Again you can use color, two colors, to denote the arrangements of these two structures with respect to this plane. Another example of color symmetry (one that is, to a certain extent, symbolic) is found in the court dances. One might say that every man or every woman interacts with his own partner, with another person's partner (as in a square dance and many other dances), and then with the other persons of the same sex. And when you try to describe the figures that the dancers create and to find their symmetry, color symmetry becomes very useful: you can say that the women are one color, and the men are another.

Some weeks ago somebody asked me: "Does color symmetry occur in nature?" I had to answer that in two ways: yes and no. If we look at a flower, we will often see that it has two, three, or four colors, but rarely—I think never—will we find, for example, a red and yellow tulip in which some petals are red and some are yellow. We might have an all-red tulip, or an all-yellow tulip, but when nature decides to mix colors, we find that the extreme ends of *each* petal may be all red while near the base each petal is yellow. We never find that in a flower equivalent points are systematically red, yellow, red, yellow. As another example, the extremities of a Siamese cat—the tips of his ears, his tail, his paws—will be black and the rest of him will be brown. This is probably because heat radiates out from any pointed or highly curved portion of a surface. These colors are functional. We would not expect that a Siamese cat will have one brown ear and one black ear.

On the other hand, in the more symbolic sense we find a great deal of color symmetry. Think of sodium chloride crystals, rock salt. As you may know, the sodium chloride structure is a sort of three dimensional checker board. Each sodium ion has six nearest neighbors of chlorine at the corners of an octahedron; and each chlorine has six neighbors (namely sodium), also at the corners of an octahedron. By classical symmetry theory we can describe the relations between the sodium atoms: they form a so-called face-centered cubic lattice. The chlorine ions, by themselves, also form this pattern. But since sodium and chlorine are two entirely different things, there is no way, in the classical symmetry theory, to express the fact that sodium is surrounded by chlorine in the same way that chlorine is surrounded by sodium. In other words, in classical symmetry theory we have no way of describing analogies; we can only express identities. As an example of another crystal structure, let us take diamond. Diamond is made of carbon atoms, each of which is surrounded by four other carbon atoms at the corners of a regular tetrahedron. Starting with a given carbon atom as center, we find four other carbons at the corners of a tetrahedron; each of the corner carbons is surrounded by the original carbon and three others, which together also make a tetrahedron. This pattern continues, so that the structure becomes an infinite pattern built of tetrahedra. Now, if we replace half the carbons by zinc, and the other half by sulfur, so that every zinc is surrounded by four sulfurs, and every sulfur by four zincs, then we have a very well-known mineral called sphalerite. Again, classical symmetry theory tells us the relations between the zinc atoms (which again form a face-centered cubic lattice) and between the sulfur atoms, but it cannot express the fact that the zincs are related to the sulfurs in a way perfectly analogous to the way carbons are related to each other in a diamond. Using color symmetry, however, we can say that the zinc sulphide structure is a two-color lattice complex entirely analogous to the monochrome diamond lattice complex.

There is a tremendous wealth of analogy in nature, and when we have analogy rather than identity, color symmetry gives us the means of expressing it.

ALICE DICKINSON

Change Ringing: Theory and Practice

At the dance festival during the Symmetry Festival, the sixteenth-century dance, the Branle de la Pois, was a visual metaphor of bells in a plain hunt. The ten dancers lined up alternately by sex. Adjacent dancers changed places, each sex starting in a different direction and alternating partners each time, staying twice at the front or end of the line. With odd-numbered males and even-numbered females, the successive positions display a central symmetry as shown below.

M_1	F_2	M_3	F_4	M_5	F_6	M_7	F_8	M_9	F_{10}	1st position
F_2	M_1	F_4	M_3	F_6	M_5	F_8	M_7	F_{10}	M_9	2nd position
F_2	F_4	M_1	F_6	M_3	F_8	M_5	F_{10}	M_7	M_9	3rd position
F_4	F_2	F_6	M_1	F_8	M_3	F_{10}	M_5	M_9	M_7	4th position

With men representing odd-numbered bells and women even-numbered bells in the tonic scale on the fundamental F_{10}, the diagram also represents tone rows produced by bells rung in a sequence of different positions. An alternative description of the movement of the dancers—or equivalently the movement of the bells in a sequence of tone rows—is given in Webster's New International Dictionary, Second Edition, Unabridged:

> *Change ringing.* The continual production, without repetition, of changes on bells. A set of bells for change ringing, called a *ring* of bells, is tuned to the diatonic scale. When struck in descending order from *treble* or highest (designated "1") to *tenor* or lowest, the bells are said to be in the position of *rounds*. *Changes* are variations from this striking order according to certain rules, as that (1) no bell shifts more than one place in a change from its position in the change preceding; (2) no change is rung twice from the time the bells leave the position of rounds until they return. The *course* of any bell is its shifting path through a series of changes, though the term may also designate the series itself. In *plain hunting* all bells work regularly from first place, or *lead*, to last, or *behind* (called *hunting up*), and back again (called *hunting,* or *coursing, down*), striking as first or last in two successive changes. A single bell with such a course is said to have a *plain hunt.*

This method of ringing is called plain bob major if there are eight bells, or plain bob *royal* if there are ten bells, as in the dance.

Why are bells rung in such a manner? In England, ringing towers are part of the landscape. Outside of England, change ringing must be explained. The casting of bells was well developed by the middle ages; eleventh-century technology produced bells weighing over a ton. Perhaps the crucial influence was a tenth-century decree in England that granted the title of Thane to a Saxon who owned five-hundred acres, a church, and a bell tower. Today, there are thousands of bell towers throughout England. As wealth became concentrated, these bell towers housed not one or two, but five or more bells weighing from a quarter ton up to two or three tons. How were the bells rung? What sort of music could be produced? On the continent bells are often rung together at random, or tunes are rung on a carillon. But in England a whole new musical system developed; it derived from the characteristics of free-swinging bells.

For centuries bells had sounded alarms, had signaled the time to get up, to go to work, to eat lunch, and had announced that the community oven was ready for baking. He who commanded the bell commanded the town. The bell was important, and everyone knew that the sound of a free-swinging bell was sweeter and would carry farther than that of a stationary bell.

The early bells were rung by a number of men pulling on a lever (fig. 1a). At Canterbury, the three-ton bell needed about twenty men; they would all strain and pull and raise it a little and then let go. The bell would swing back, the clapper would hit and strike a note. For greater control, a wooden half-circle with a rope, was attached to the head stock (fig. 1b). The pull of the rope turned the half-wheel and the bell swung with the half-wheel. Control was minimal but bells could be rung in order. The first notes would be separated; then sounds would jangle as the bells swung back and forth.

The definitive change that allowed real control was a full wheel attached to the headstock (fig. 1c). Then, if the bell is mouth up when the rope is pulled, the bell will fall and swing through a full 360°. The clapper will rest as the bell falls, but as the bell rises on the other side, the clapper falls and hits the lower edge. Thus a single note sounds. This is the present method of bell ringing.

A swinging bell has properties in common with the pendulum or yo-yo. The timing cannot be changed. You cannot make a pendulum of fixed length swing faster. The rate at which the pendulum swings, or the yo-yo goes up and down, or the bell swings through 360° are all fixed; there is no way to hasten the process. Since each bell rings at a fixed rhythm

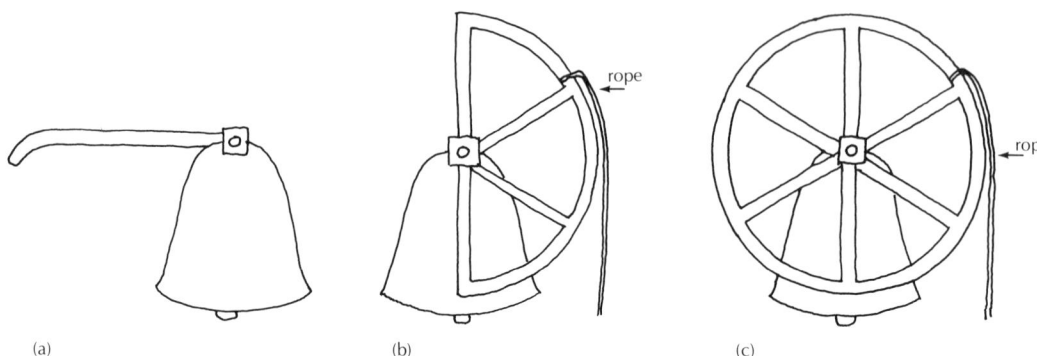

1. (a) An early bell, rung by pulling on a long lever. (b) A bell, attached to a half-wheel, rung by pulling a rope fastened high on the wheel. (c) A bell attached to a full wheel. This mounting, presently used in change ringing, permits a single note to be sounded.

which cannot be changed, familiar tunes cannot be rung. However, eight bells on the tonic scale can be pulled off one after another, to sound the eight notes in the descending scale, and then the first and following bells may be swung back in the other direction to sound a like sequence of eight notes. Thus, the descending scale can be repeated over and over at a fixed rhythm. This repeating sequence, a lovely cascading sound, accompanies English brides and grooms down the aisle after their wedding vows.

But even this exciting sound dulls with repetition. Once a ringer has learned to handle a rope well she can pull the rope so that the bell swings through 360° and balances mouth up. The ringer whose bell follows immediately after can give the bell a lighter pull than usual so that it doesn't quite get up to the balance point, and therefore comes down and rings sooner on the back stroke. Thus two bells can change places in the tone row. In particular, if the person ringing the number one or treble bell (lightest and highest pitch) could attain the balance position and hold it while the number two bell is pulled down ahead of it, then bells one and two would have changed positions. If each adjacent pair of bells changed places, then the sequence 1 2 3 4 5 6 7 8 would be followed by 2 1 4 3 6 5 8 7.

It is not practical for a bell to change more than one position at a time. Balancing a bell is always tentative, and if the ringer moving ahead stopped the bell too far below the balance point, the extra rope released would flap about in the ringing chamber, and the bell would not rise sufficiently on the other side. Or if the rope is pulled too hard—to get the bell mouth up again—the bell might swing over the top and the ringer might be pulled to the ceiling. (The first principle in ringing is, if your feet leave the floor, drop the rope and fall!)

The basic rules of ringing are: (1) the rhythm is regular; (2) each bell sounds once in each tone row; (3) ringing always starts and ends in rounds; (4) each bell can move only one position at a time; (5) no tone row, or sequence may be repeated (except for rounds at the beginning and end of the composition).

An example is a lead of plain bob minor (minor indicates 6 bells). All the bells are *hunting*. The changes in position of two adjacent bells is called a transposition. The particular method shown here alternates between three transpositions and two transpositions, from row to row.

Since rounds terminate any composition, a longer touch of the same method can be rung only if the last change is other than rounds. For example, this *lead end* of plain bob minor could be extended to a *plain course* of bob minor, if the last change listed were 1 3 5 2 6 4, as shown below, and then the bells continued to hunt. (In most compositions, when the treble bell leads or is first in the tone row, each bell executes an unusual pattern, often called a *dodge,* to prevent repetition of a previous tone row. This allows another distinct sequence of tone rows to be added to the composition.)

Plain Bob Minor

```
1 2 3 4 5 6
2 1 4 3 6 5
2 4 1 6 3 5
4 2 6 1 5 3
4 6 2 5 1 3
6 4 5 2 3 1
6 5 4 3 2 1
5 6 3 4 1 2
5 3 6 1 4 2
3 5 1 6 2 4
3 1 5 2 6 4
1 3 2 5 4 6
1 2 3 4 5 6
```

In the diagram below, the changes in column one are followed by those in column two, and so forth (fill them in). You can see that the alternating transpositions are the same in each column, and that the same dodge is executed at the end of each of the five columns to avoid rounds.

Plain Bob Minor

1 2 3 4 5 6				
2 1 4 3 6 5	3 1 2 5 4 6	5 1 3 6 2 4	6 1 5 4 3 2	4 1 6 2 5 3
2 4 1 6 3 5	3 2 1 4 5 6			
4 2 6 1 5 3	2 3 4 1 6 5			
4 6 2 5 1 3	2 4 3 6 1 5			
6 4 5 2 3 1	4 2 6 3 5 1			
6 5 4 3 2 1	4 6 2 5 3 1			
5 6 3 4 1 2	6 4 5 2 1 3			
5 3 6 1 4 2	6 5 4 1 2 3			
3 5 1 6 2 4	5 6 1 4 3 2			
3 1 5 2 6 4	5 1 6 3 4 2			2 1 3 4 5 6
1 3 2 5 4 6	1 5 3 6 2 4	1 6 5 4 3 2	1 4 6 2 5 3	1 2 4 3 6 5 *dodge*
1 3 5 2 6 4	1 5 6 3 4 2	1 6 4 5 2 3	1 4 2 6 3 5	1 2 3 4 5 6

In addition to the musical sound of change ringing, the art has engendered a new language because all communication while ringing must be brief and precise. (There is a published Glossary composed of eighty-three pages of technical terms.) The flavor of the exercise is well conveyed in Dorothy L. Sayer's book *The Nine Taylors*.

To ring all possible changes on eight bells would take twenty-four hours; on seven bells, the possible 5040 changes would take three and a half hours. These 5040 changes, called a peal, are frequently rung; this is always done without any written music. Indeed when a ringer is handling a bell rope, watching other ropes that are to be followed or preceded, and at the same time listening to the music, there is no possibility of reading a score. How do ringers remember and execute such a lengthy composition? What is the trick? As you might guess, the most frequent method is to rely on symmetry. For example, the line in figure 2 indicates the symmetric path that a bell follows in a particular composition.

Change ringing has flourished since Fabian Stedman published his *Tintinologie* in 1668. Paul Revere was a ringer who also cast bells. Indeed, he kept the key to the belfry in the Old North Church, and thus, perhaps, bell ringing influenced our history. Belfries are traditionally locked because novices, climbing among heavy bells that turn easily, can

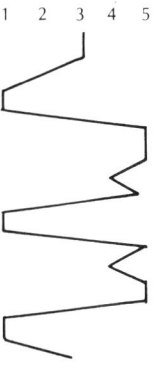

2. Path of Bell 3 in a Plain Course of *Grandsire Doubles*.

be crushed. In most towers, the bells are always lowered, that is, left in a mouth-down position, before ringers leave and lock the tower.

In the same period Lagrange (1736–1813) in France was developing a mathematical theory of permutation groups of which change ringing compositions are examples, but the theory and the examples were developed in different countries.

Some mathematicians find a special delight in change ringing. Troyte, a great ringer of the last century wrote, "On all numbers of bells exactly half the changes are of one nature, and half of another; what this nature is, it is out of my power to explain, but as will be seen by-and-by it is a *fact* which must be understood before it is possible to go into the science of composing and proving peals."[1] It is now known that half the changes may be obtained only by an even number of transpositions, and the other half by an odd number of transpositions. The study of these permutations has a central place in the mathematical theory of groups, and some change ringing compositions raise unsolved problems in mathematics. How a nonmathematical ringer sensed the difference in these changes is difficult to understand, but the difference is indeed crucial in both ringing and in composing, both of which are ancient and active arts. Thus change ringing continues to delight the ear and the mind.

ELLIOT OFFNER

**Renaissance Typographic Ornament:
Origins, Use, and Experiments**

1. A Nuremberg arabesque.

This discussion is about a very small aspect of printing and typography, namely the decoration of books with small units of printing types upon whose faces decorative bits of patterns have been cast. These decorative type units can be placed in a variety of arrangements and when inked and impressed into the paper they will leave their design. Such a unit looks like a piece of type, except that instead of a letter upon the face, a design such as the one in figure 1 may appear. This arabesque dates from 1721 in Nuremberg.

My interest in these ornaments derives from my work as a printer and a calligrapher, and my fascination with the inherent symmetrical design possibilities of these wondrous bits of metal. I can add little to scholarship in this field; for me the temptations of substantiating one's hunches and observations with thorough, time-consuming research give way to the inevitable retreat to my studio. But the published material itself is somewhat obscurely located, appears always in limited editions, and is very difficult to pull together. Few monographs exist on the extant ancient typographical stores of the early printers and typefounders. Much is conjecture. Perhaps the most important single contribution to modern scholarship in ascertaining the origins of early metal typographic ornaments was the publication in 1967 of *John Fell: The University Press and the 'Fell' Types,* by Stanley Morison with the assistance of Harry Carter. Here Morison's own pioneer work, "Printers Flowers and Arabesques" in the first issue of the *Fleuron* in 1923, is revised and updated. Fell, who was delegate of the Oxford University Press, Dean of Christ Church, Vice Chancellor of the University, and Bishop of Oxford, bequeathed to the University a mass of typographical material bought mostly from Plantin and used at the Press at Oxford. Morison has catalogued all of this ancient material, a large amount of which seems to have been cast from matrices sold to Plantin by Robert Granjon, the talented French punch-cutter, whose name is central to this discussion. This book itself is out of print and its predecessor, *Notes on a Century of Typography at the University Press, Oxford, 1693–1794,* by Horace Hart in 1900, was published in an edition of 150 copies and, until a reprint in 1970, was rarely to be seen. Hart was printer to the University from 1883 to 1915, and it was he who made the first exact study of the origins and history of the mass of ancient typographical material that came into his charge in a state of total disarray. Figure 2 is the dust jacket from *John Fell,* showing a pattern derived from left- and right-hand arabesque units, which are of considerable historical importance as they are probably part of Granjon's famous twelve-piece fleuron which I shall discuss later. Only one matrix for this ornament survived; in 1895 the matching matrix was made and the type cast, and the units have been in use since then at Oxford. Hart shows the ornaments in the *Century* and cites 1609 as the first usage in England.

Type ornament of this sort, which is arabesque in origin, actually arrived in its mature state by the middle of the sixteenth century in France. How the arabesque came into use principally in France at first, and why it did not gain wide acceptance in other places—Rome for example—are complex questions. Metal type ornaments were handsome inventions (probably by Granjon), and they played a part in securing the lead that France wrested from Italy in Renaissance typography in the mid-sixteenth century. The type ornament unit has proved durable as well as revolutionary, surviving to our time and not yet pronounced dead. While it seems to be passing out of trade use, it remains in the hands of artists and compositors who continue to explore the potential for new pattern design. It is from artists that we can expect to see some remarkable new results that have been hinted at by past designers. It is true that type ornaments were conceived originally as accompaniment for the printed word, but by 1742 Simon-Pierre Fournier's specimen book (fig. 3) displayed ornaments of this design in flamboyant arrangements that suggest design for its own sake, a departure from past practice.

To understand the origins of the earliest flower type units, or fleurons, we must review some of the background. In Venice the earliest typographic books were printed with spaces left for the handpainting of initial letters, engraved or painted borders, painted paragraph marks, and so forth. Hand decorative work was tied up with the purchaser and book seller, and the book was therefore not a finished product. But very soon publishers seeking to market a more finished product began to issue books with wood-engraved initial letters, ornamental head and tail pieces, and by 1476 there evolved two elements never known to the medieval manuscript maker: the title page and the printer's mark. The title page eventually became the main arena in which the artist-compositor practiced his skill in the manipulation of printers' flowers into handsome designs and borders. The credit for the first title page must go to that most remarkable printer of Augsburg, Erhardt Ratdolt.[1] Figure 4 shows Ratdolt's title page

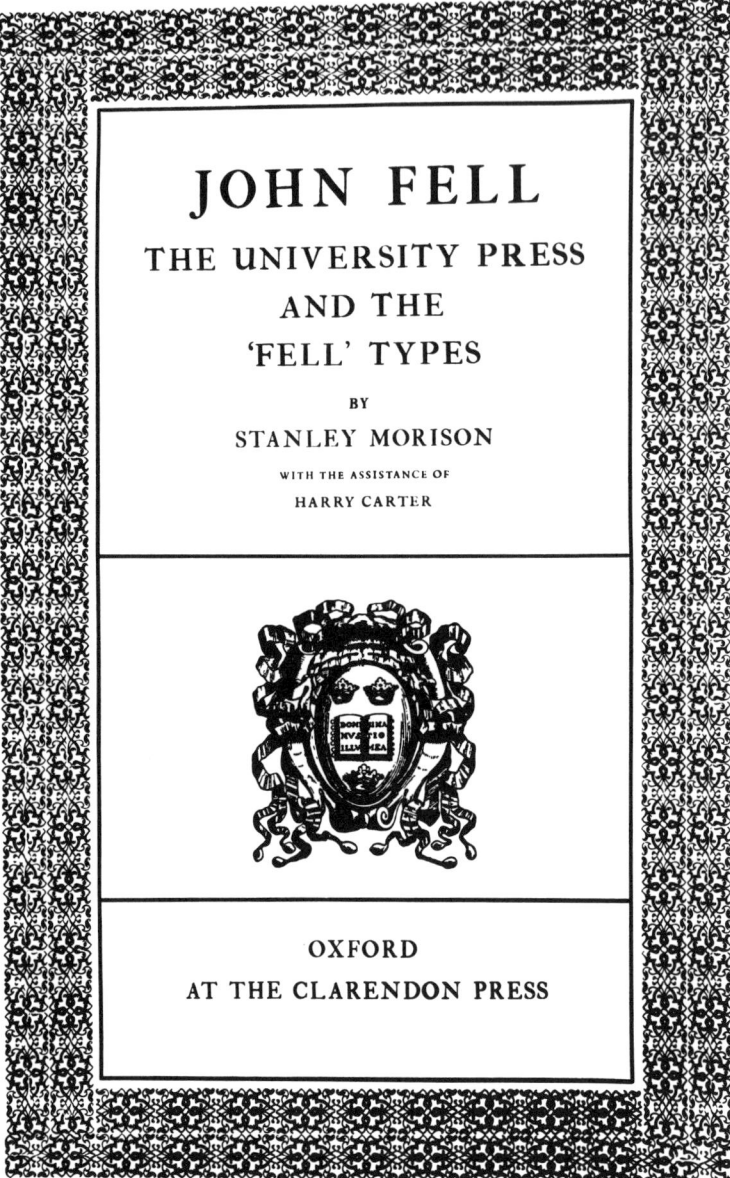

2. Dust jacket from Stanley Morison, *John Fell: The University Press and the 'Fell' Types* (Oxford: The Clarendon Press, 1967). Used by permission.

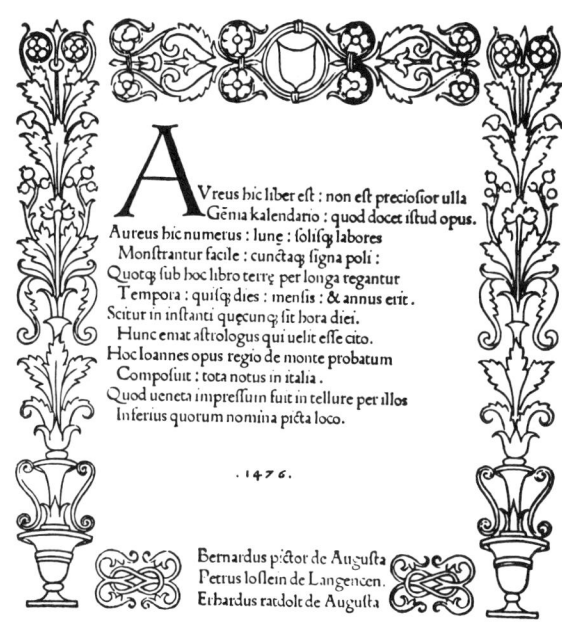

3. (Left) Type specimen book of Simon-Pierre Fournier, Paris, 1742. 4. (Right) Jo. Regiomontanus, *Calendarium* (Venice: Ratdolt, Maler and Löslein, 1476).

for a fifty-five year calendar calculated by the German astronomer, Johannes Regiomontanus. The light and beautiful woodcut border (cut by Bernard Maler) was the forerunner of great numbers of title borders through five centuries. (For a more comprehensive discussion of title pages, I refer you to A. F. Hohnson's essay *Title Pages, Their Forms and Development,* London, 1928; and E. Ph. Goldschmidt, *The Printed Book of the Renaissance,* Cambridge, 1950.)

Yet it is not clear how much the needs of the title page aided the development of the fleuron, for it took a long time before type ornaments found their way into the title pages. The printers' flower appears to have come from the small tools known as *piccoli ferri,* or little irons, used to stamp binding. The technique is still used by hand binders.

If the arabesque is a key form in Renaissance printers' flowers, there is no question that the vine leaf was understood to be a part of the arabesque. We shall see that more clearly as we go on. The leaf, however, is by no means unique to the arabesque. The vine leaf in Graeco-Roman sources was used in a variety of ways, including as a form of punctuation epigraphically and in manuscripts. Alison Frantz's "Byzantine Illuminated Ornament" is one source where many of these forms are traced.[2] How great was the dependence upon Graeco-Roman sources for original

5. Binding of the *Kuran of Oeldjaitu*, A.D. 1313, ms., Bibliothèque Khédivale, Cairo, reproduced from Bernhard Moritz, ed., *Arabic Palaeography* (Leipzig: Hiersemann, 1905).

Moslem ornament is another question which needs more exploration.

Figure 5 shows the binding of a fourteenth-century manuscript in the Kedivial Library, Cairo. It gives us a good view of the sort of arabesque pattern which apparently first came to the attention of the early Venetian printers and ultimately played such an important part in the book arts. Variations on these patterns reappear in the lace pattern books and other craftsmen's books of the sixteenth century.

The chart in figure 6, with its eight ornaments, is from the Morison-Meynell article in no. 1 of *The Fleuron* and is a convincing presentation of the relationship between orthodox Islamic arabesque design and the early arabesque typographical flowers. The diagram also affords the opportunity to see how these units break down into component parts (figs. 6d and 6e) and how the vine leaf is incorporated into the arabesque. Equally interesting is the series of elements (fig. 7) redrawn from arabesque borders used by Jean de Tournes after the mid-sixteenth century.

6. Ornaments from Stanley Morison and Francis Meyhell, "Printers' Flowers and Arabesques," *The Fleuron: A Journal of Typography*, no. 1, p. 6, London, 1923. (a) Diagram of one of the most common forms of Arabian and Persian architectural ornament, tenth century (from Charvet, Arts Decoratifs, Paris, n.d., p. 275). (b) A twelfth-century oriental calligrapher's ornament, whose outline has correspondence with that of figure 6a (redrawn from B. Moritz, *Arabic Palaeography,* Leipzig: Hiersemann, 1905). (c) Element from a French renaissance woodcut title-page design (Jean de Turnes, Lyons, 1557), perhaps designed by Bernard Salomon. (d) and (e) Enlarged reproductions of the earliest arabesque flowers. Cast of four metal units, they combine to produce the outline figure resembling figures 6a and 6b. These two flowers occur not later than 1557, and are probably of Lyons or Antwerp provenance. (f) Another variety, founded in two pieces. A similar motif is to be found in a woodcut fleuron used by the Trechsels, Lyons, 1540 (*Baudrier,* XII, 280 *bis*). (g) An enlarged unit of figure 6e, with an additional leaf. (h) A simpler form of figure 6g. This flower is that most frequently found in current printing.

7. Elements redrawn from borders used by Jean de Tournes (Lyons, 1557). From Stanley Morison and Francis Meyhell, "Printers' Flowers and Arabesques," *The Fleuron: A Journal of Typography*, no. 1, p. 7, London, 1923.

The very first ornaments cast on type bodies are shown in figure 8, taken from a book printed by the brothers Giovanni and Alberto Alvise in Verona in 1478. Crudely cut and of questionable merit aesthetically, they appear long before cast type ornaments came into general use. Morison refers to the Verona work as having neither ancestry nor progeny and dismisses it. The Verona scholar and printer Dr. Giovanni Mardersteig, however, has been engaged in an examination of Verona Incunabula, particularly the work of the Alvises, and we may learn some new things shortly. Mardersteig's own press, the Officina Bodoni, has published what he calls his most important work, *The Fables of Aesop,* which included recuttings of the Alvise ornaments used in the *Aesop* of 1479.[3]

Before we go further in our consideration of type unit ornament, a word may be necessary about the main body of book ornament, the technique of which was the woodcut border. It was classical, architectural, and floral, and appropriately inspired by the antique. A key figure in its earliest development in Venice again is Ratdolt; his 1476 *Calendarium* (fig. 4) has a fine light four-piece border, each side with a classical vase from which a symmetrical cursive floral decoration rises. The floral decoration is mainly trefoil.

To understand the development of typographical ornaments in the second quarter of the sixteenth century in France, we must realize that Italy provided the impetus for the new intellectualism: Frenchmen studied in Italy, and Italian craftsmen, at the invitation of Francis I, settled in France. The French printers quickly absorbed the lessons of their Italian mentors (the influence of Ratdolt was a strong one), and gradually there evolved a distinctive mode of book production that was quite French and very advanced. The scholar-printers of Italy were being equaled in scope and ability by their French protégés.

Geofroy Tory is the most frequently mentioned artist in the earliest stages of the new book design in France. His influence on decoration affected the whole concept of book design. After a stay of uncertain duration in Italy, he returned to Paris about 1525. The lightness of his borders, which was like Ratdolt in spirit if different in personal style, and their compatibility with the lighter French type brought about a new harmony of the whole book. The Gothic heaviness of Urs Graf, Holbein, and the South German school gave way. Tory's work was intertwined with that of the artist engraver Simon de Colines and paralleled that of the Estiennes who, among others, reflected earlier Italian books. But I think we can point to Tory as the model of a major figure who led the Italian Renaissance.

RVBRICA

ALNOME DEL NOSTRO
SIGNORE
IESV CHRISTO
E DELA
SVA GLORIOSA MADRE
SEMPRE
VERZENE MARIA
COMENCIA VNO BELLO
TRACTATO
ALA CREATVRA MOLTO
VTILE
ET ANCI NECCESSARIO
CIOE
DELA SCIENTIA ET
ARTE
DE BEN MORIRE
ET BEN
FENIRE LA VITA
SVA

8. Carpanica, *Arte de ben Morire*, Giovanni and Alberto Alvise, Verona, 1478.

Yet it is always important to remember that ideas and opinions were in competition, and orderly chronological development is not easy to establish. The intellectual fervor and curiosity over stimulating artistic ideas created an atmosphere in which splendid new developments could occur almost without warning; ideas which seemed spent could reappear with new vigor in modified form. And in the first quarter of the sixteenth century in Italy, the arabesque nearly vanished—the one notable known exception is a Christian horary written in Arabic. Interlace, on the other hand, first used by Feliciano in 1475 in a Petrarch (fig. 9), suffered no decline. Morison suggests this may have been because of its Christian as well as Levantine origins.

But in 1527 the gifted Venetian calligrapher Giovanni Antonio Tagliente published his *Essempio di Recammi,* an embroidery pattern book. His was not the first one: a few German model books preceded him by several years, and he must have seen one or more of them, as well as the sheets of decorations used by Venetian craftsmen and craftswomen. However, Tagliente's book included as its major section arabesque and moresque designs not seen in the others. Esther Potter[4] emphasizes the importance of Tagliente's *Essempio* as the forerunner of a line of such pattern books and as an important source on the development of Italian Renaissance needlepoint. Figure 10 is a redrawing of one of Tagliente's originals and is taken from Potter's article. I think it shows arabesque structure even more clearly than the original. Few if any of Tagliente's designs had been used in book design before he published his book.

The proliferation of pattern books was great, and after Tagliente they all contained the arabesque. Lyons was the city where their production was greatest and where the arabesque was to become a major feature of book decoration. The Lyonnese considered Peter Flötner's book of 1546, which was produced in Zurich, their finest model (fig. 11). One troublesome question has always been, Why did we not see the wide application of arabesque typographically by mid century? The speculation has been first, that the antique revival was a stronger force, and second, that there might have been problems involved in using the orthodox arabesque identified with the Turks, enemies of Christendom. Being more exalted than swordhilts, fabrics, and the like, books remained free of these designs. As Lyons was emerging as the great center of French printing, a dominating number of artists and printers there were Protestant and therefore not unwilling to use the arabesque of the pope's enemies, the Turks.

From Flötner's model book came designs for many printers, including Jean de Tournes and his extraordinary collaborator, Bernard Salomon. Salomon cut most of the single-piece engravings appearing in de

9. Woodcut frame from a *Petrarch* printed by Felice Feliciano, Verona, 1475.

59 Renaissance Typographic Ornament

10. Redrawing of ornament from Giovanni Antonio Tagliente, *Essempio di recammi* (Venice, 1524), reproduced from Stanley Morison, *Splendour of Ornament* (London: Lion and Unicorn Press, Royal College of Art, 1968), p. 10. Used by permission.

11. Metal etching designs from Peter Flötner's pattern book, Zurich, 1546.

12. Ornaments, probably by Bernard Salomon, from the print shop of Jean de Tournes, Lyons, 1558.

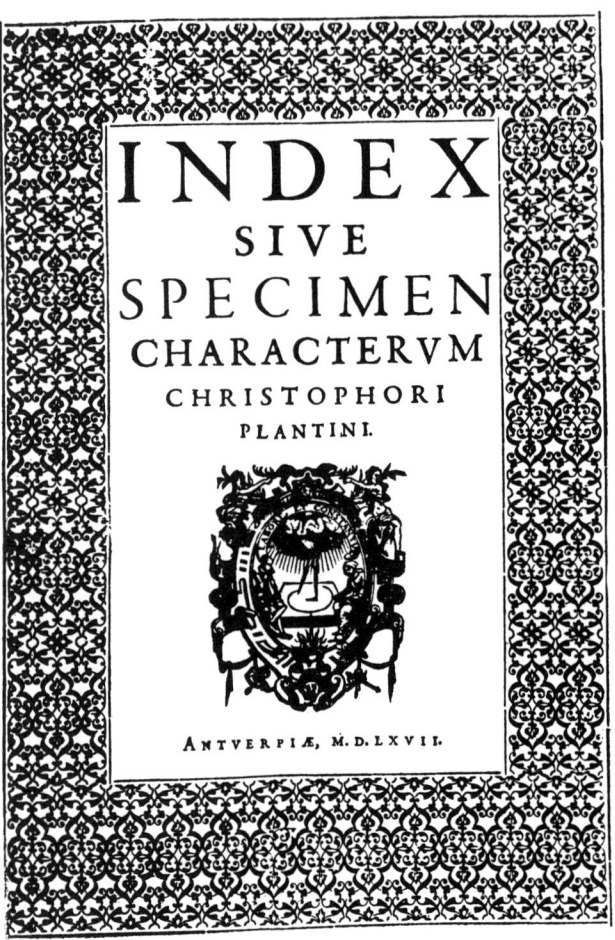

13. Title page of Christophori Plantini, *Index Sive Specimen Characterum* (Antwerp, 1567), from Hendrik O. L. Vervliet and Harry Carter, *Type Specimen Facsimiles II* (London: The Bodley Head, 1972). Used by permission.

Tournes's books. Figure 12 shows a group of his engraved head and tail pieces and borders. Salomon was preoccupied with the intricacies of the arabesque throughout his lifetime, and he created many designs for silks, jewelry, and furniture as well.

The next development is the one with which we are most concerned: the mature arrival of the cast metal type ornament. The fleuron gave to every compositor with a reasonable sense of symmetry the opportunity to embellish the printed word, to create easily a design which would please the eye.

Although it has not been proved conclusively, all indications are that it was Robert Granjon, son-in-law to Bernard Salomon, who brought the invention to that state which has not been surpassed. Other artists had cut out sections of arabesques and used them singly or in strips on bindings. Tagliente's pattern book actually shows a less sophisticated usage of this technique. Nevertheless, whatever events preceded this new product of the typefounder, we must ascribe the invention of functional type unit flowers to a period probably a few years after 1550. To Granjon we must give at least the major credit for the spread of these flowers.

In figure 13 are the units on the title page of Plantin's *Index Characterum*, about which we will say more in a moment. What was unique was that the arabesque had been perceived as it had not been perceived before. Through this perception the structure of the arabesque was revealed and the component parts recognized and placed upon type bodies. It strikes me that the form of intellectual inquiry that led Felice Feliciano, Fra Luca Pacioli, Geofroy Tory, and others to examine in detail ancient epigraphic inscriptions and to seek the underlying geometric principles of the Roman majuscule was the same form of intellectual inquiry that led Granjon to seek to understand the underlying principle of construction of the arabesque.

Of further significance was the discovery of the principle of varied combinability of the units. With a single unit, or two perfectly symmetrical units, astonishing possibilities exist. The particular combinations in figure 14 were arranged by the contemporary English typographer, John Ryder, who was also responsible for the revival of these units. Ryder's work on printers' flowers is particularly important because he gives us historical information as well as new possibilities for typographic use. It is quite simply a matter of turning the units in different ways, adding and subtracting, and so forth. So skillfully worked out was the geometric basis of these early designs that the pairs could be grouped with other pairs or single pieces making trefoils, quatrefoils, and other complex configurations.

Plantin's *Index Characterum* was the first type specimen to appear in

14. A page from Christophori Plantini, *Index Sive Specimen Characterum*, op. cit. Used by permission.

15. An arabesque strip from Plantini, *Index Sive Specimen Characterum,* op. cit. Used by permission.

book form. A key document in the study of ancient type material, it has recently been republished in facsimile with an introduction by John Dreyfus. We know now that the ornaments shown here were sold to Plantin by Granjon (some of them are shown in figure 6). Inside the specimen we see two strips using the same ornaments but achieving slightly different effects.

But the most unusual arabesque strip is shown in figure 15. Plantin's not-so-perfect letter press printing does not help in visually dissecting this ornamental strip. We know that the matrices were purchased from Granjon and it is a twelve-piece fleuron of some fame. Marcus Van Voernewyck, a contemporary of Granjon, wrote a rhymed chronicle of the history of Belgium. One of the verses is about Granjon and his miraculous twelve-piece fleuron. The border from the John Fell book (fig. 2) is a relic from that fleuron.

The dynamics of the method were accepted widely, and the border, head, or tailpiece became one of the conventions of printing. Each shop with a supply of fleurons could construct different designs for its books. This durable contribution to the art of printing was well used, abused, forgotten, and revived.

Figure 16 is a copy of the *Perambulation of Kent* printed in London using Granjon ornaments that had also appeared in Plantin's *Hadriani* of 1565. We recognize the units in the center above the imprint as being from the *Index Characterum* (fig. 13). Johann Mayer at Dillengen printed the *Ritus Ecclesiastici Augustensus* (fig. 17) in 1580. The border uses multiples of twelve different sorts.

The handsome 1577 border by the Lyons printer Guillaume Rouille shown in figure 18 is made up of a six-unit set cut by Granjon that has received wide usage because of its astonishing versatility. Not only does it survive today, but it has been the subject of at least three twentieth-century published exercises that show its incredible capacity to achieve new combinations and afford the artist further expression. Its component parts are displayed in figure 19. In figure 20 the six-piece set seems to be used again. But on closer examination it is revealed that the process has been reversed. It is a one-piece fleuron, apparently engraved on a block. After an impression was taken from the built-up design, it was transferred to the block where it was cut as one piece. Edward Allde was the printer in 1610 in London.

By now the typefounder's art was dedicated to the production of the fleuron as well as printing types, and the convention of constructed borders and decorations was universal. But the inevitable changes in art and architectural styles were to have their effect upon the aesthetic

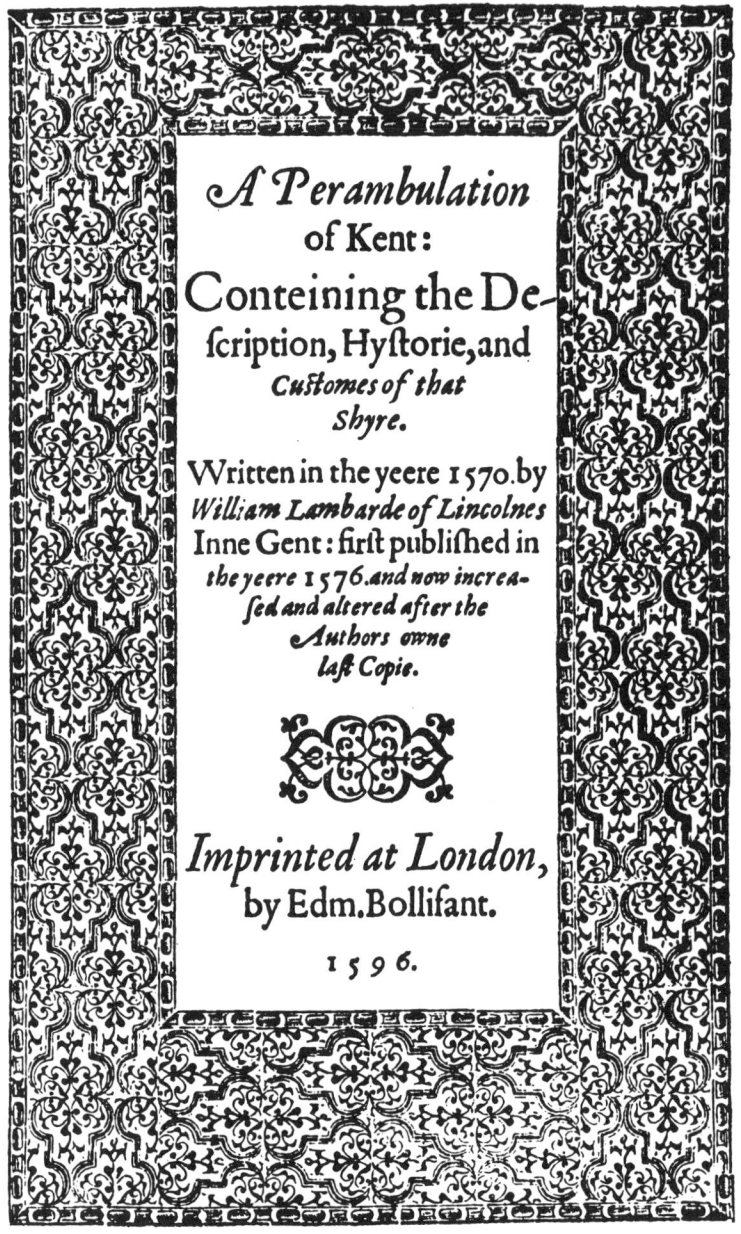

16. Title page of William Lombard, *A Perambulation of Kent* (London: Edm. Bollifant, 1596).

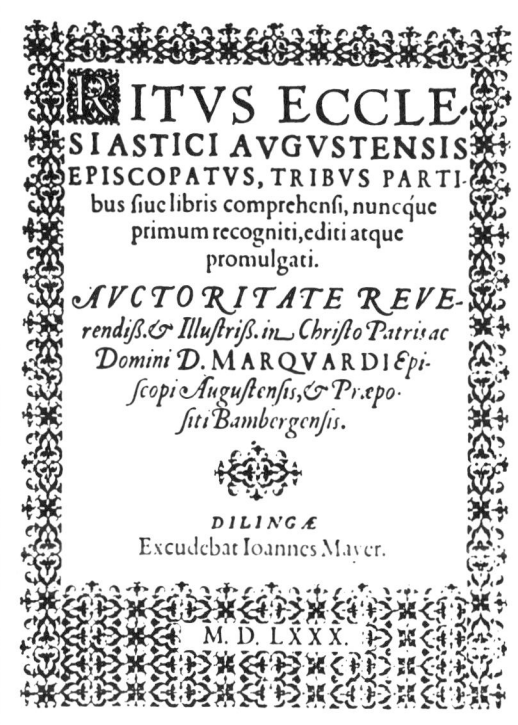

17. Two pages from *Ritus Ecclesiastici Augustensus,* Johann Mayer, Dillingen, 1580. Reproduced by permission from *The Typographic Book, 1450–1935,* by Stanley Morison and Kenneth Day (University of Chicago Press, 1963 and Ernest Benn Ltd., London). Copyright © 1963 by Stanley Morison and Kenneth Day.

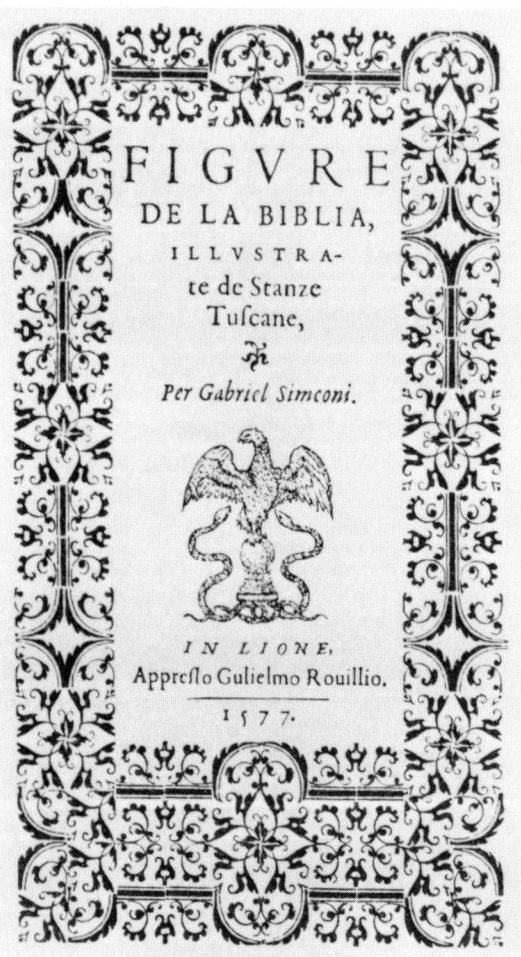

19. Six-unit arabesque of Robert Granjon.

20. Woodcut vignette (London: Edward Allde, 1610).

18. Title page from Gabriel Simconi, *Figure de la Biblia*, Guillaume Rouille, Lyons, 1577. The border is constructed from the six-unit set shown in figure 19. Reproduced by permission from *The Typographic Book, 1450–1935*, by Stanley Morison and Kenneth Day, op. cit.

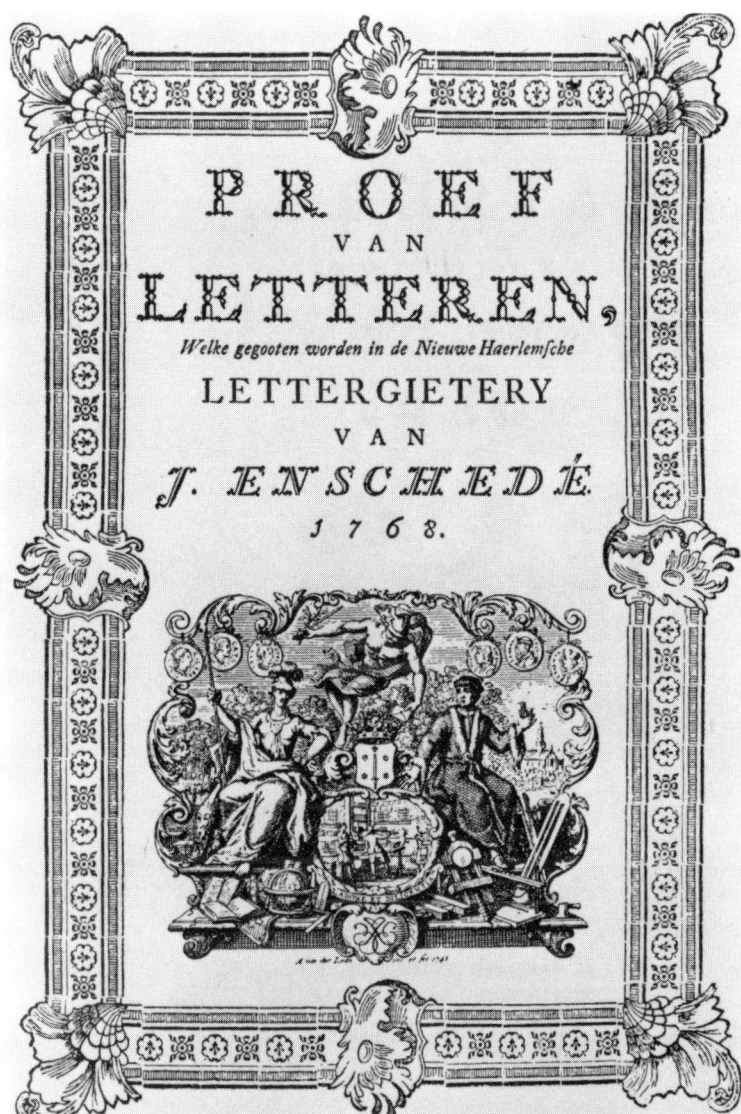

21. Type specimen of the House of Enschedé, Haarlem, 1768. Reproduced by permission from *The Typographic Book, 1450–1935,* by Stanley Morison and Kenneth Day, op. cit.

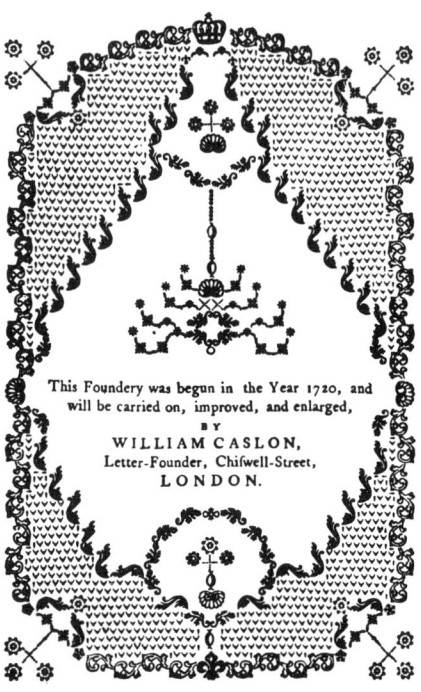

22. Colophon from the specimen book of William Caslon, letterfounder, 1785.

sensibilities of artists and printers. It is a tribute to the genius of Granjon that the construction sytem of type fleurons did not change. There was something practical in the system as well as something fundamental to man's need to build harmonious symmeterical patterns.

When Pierre-Simon Fournier published his specimen book in 1742, it was filled with rococo fleurons of all sizes and shapes creating effects very much of its own time. Actually, in many cases his designs were not original. He borrowed heavily from Louis Luce who had neither the skill with the burin nor the same fine taste of Fournier. Because of Fournier's capacity to refine and combine, he enjoys the greater reputation by far. Fournier's powerful influence can be seen at the House of Enschedé in their elegant specimen of 1768 (fig. 21), and across the channel in William Caslon's specimen of 1763, a daring design for England (fig. 22).

By the mid-nineteenth century, founders were issuing elaborate specimens in which literally thousands of pieces were put together to form bizarre and fantastic edifices. The unidentified sheet in figure 23 was typical, but by no means the best of these. Modern scholarship has only begun to take account of the large body of material, largely unknown

23. Unidentified type specimen sheet, nineteenth century.

24. A page from Stanley Morison, *"Monotype" Flower Decorations: Fine Ornament and Decorative Material Available to "Monotype" Users* (London: Lanston Monotype Corporation Ltd., 1924). Reproduced by kind permission of The Monotype Corporation Ltd.

except to a few specialists, because these creations had little use beyond their presentation in specimen books. Space prohibits me from discussing the work of many other important figures involved in this scholarship, but the Whittinghaus at the Chiswick Press (founded 1789), the Rev. C.H.D. Daniel, late provost of Worcester College, Oxford, and of course, William Morris should be recognized in this chronology.

By 1925 a serious resurgence began in good printing and in the sensible yet imaginative use of printers' flowers. In 1923 Morison and Meynell published their article on printers' flowers and arabesques, one year after the Harvard Press had issued Daniel Berkeley Updike's *Printing Types: Their History, Forms, and Use.* Updike's Merrymount Press was setting the standard for the trade while the Monotype Corporation, at Morison's direction, was recutting historic types and flowers for trade use. Thus people of talent were given the means to develop a little renaissance, the last vestiges of which may now be suffocating under the combined weight of unmastered technology and ignorance.

In 1924 a quarto entitled *"Monotype" Flower Decorations* was published bearing Morison's initials. Figure 24 is one page of a dozen which

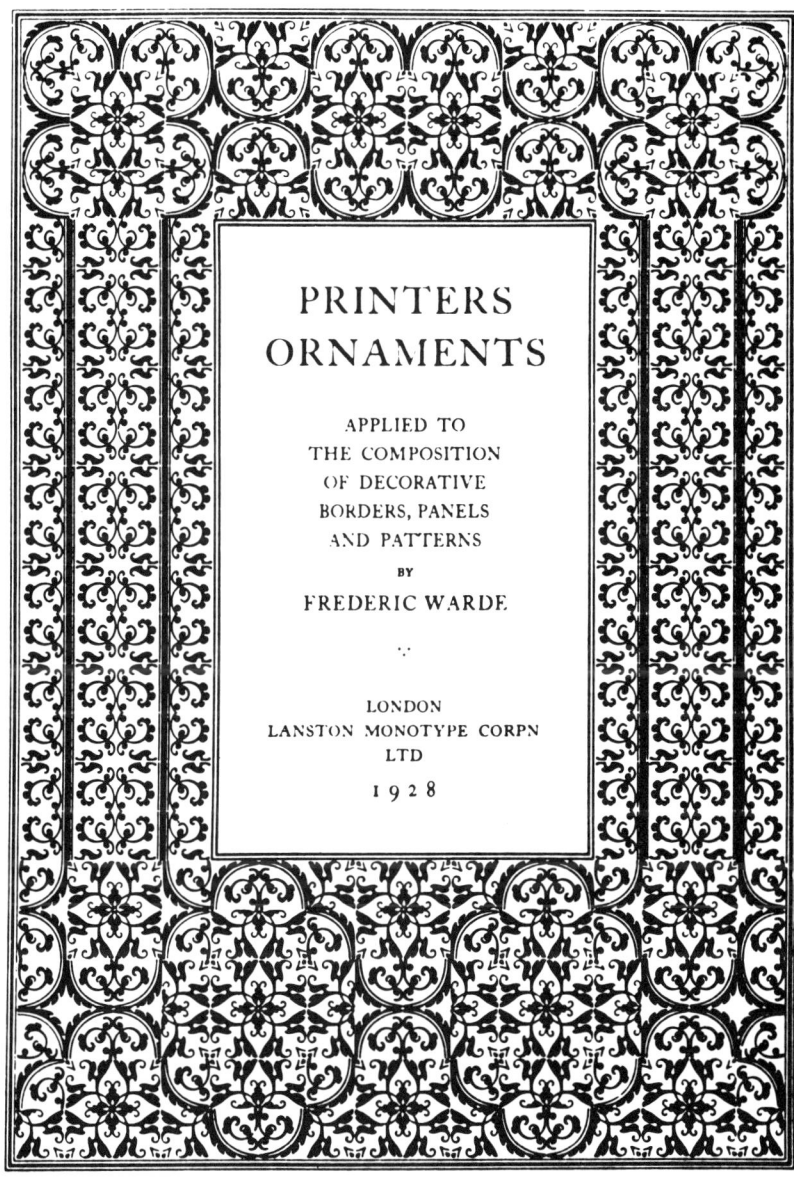

25. Title page of Frederic Warde, *Printers Ornaments,* London: Lanston Monotype Corporation Ltd, 1928. Reproduced by kind permission of The Monotype Corporation Ltd.

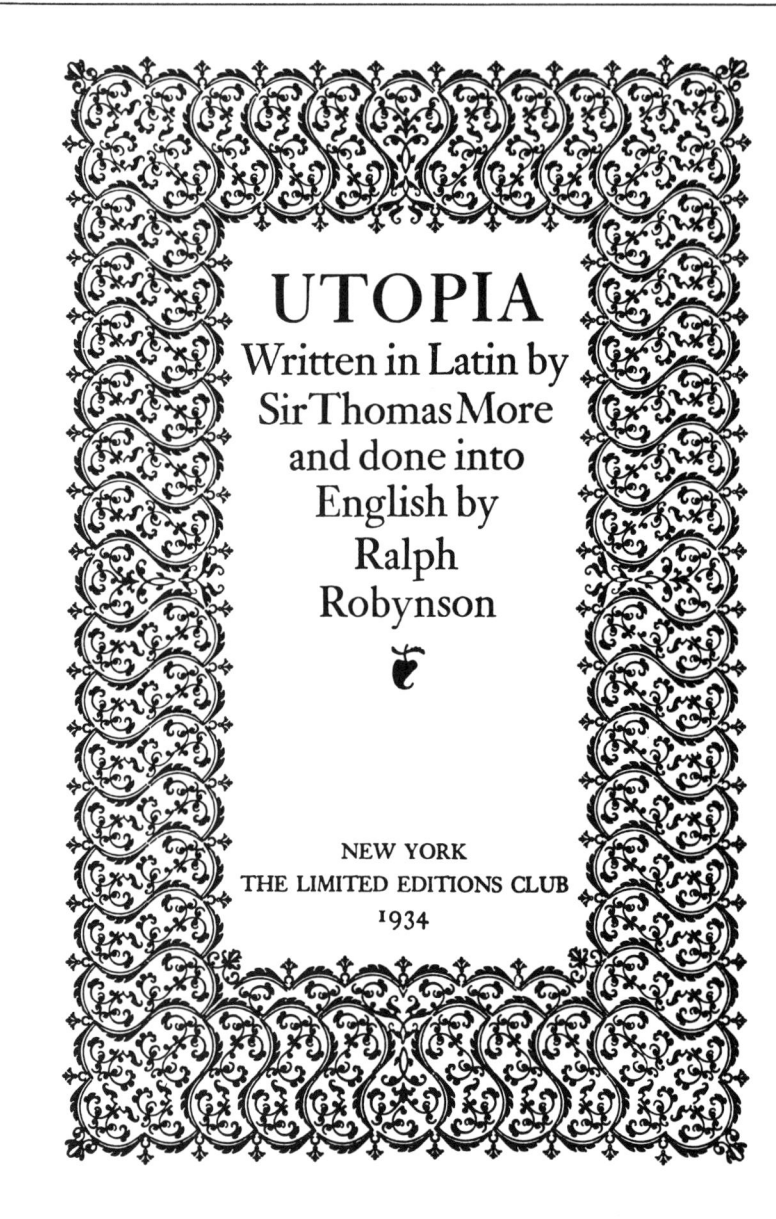

26. Title page of Sir Thomas More, *Utopia*. Copyright © 1935, 1963 by The Limited Editions Club, Westport, Conn. Reproduced by permission.

Renaissance Typographic Ornament

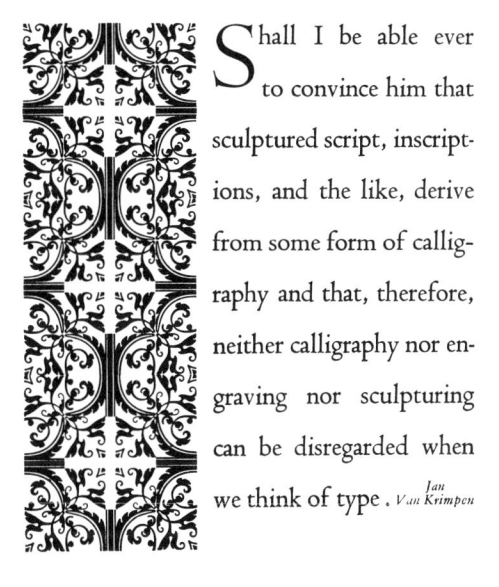

27. An arrangement of ornaments with type, from Elliot Offner,
The Granjon Arabesque, Northampton: The Rosemary Press, 1969, p. 16.

display recut decorative material in various tasteful arrangements. Most of the flowers are of Renaissance provenance and were lively proof that there were still undiscovered properties after more than 350 years. In 1928 Monotype issued another publication of greater scope called *Printers Ornaments*. This time the arrangements were by Frederic Warde, the talented American printer and sometime director of book design at the Princeton University Press. In figure 25 we see the familiar six-piece Granjon fleuron.

Perhaps the most skillful artist of this century whose life was devoted to the design of books was the great gentleman from Indiana, Bruce Rogers. No one who has handled books or type with care can fail to appreciate the extraordinary skill and personality which mark everything he designed. The pieces used in the *Utopia* title page (fig. 26) are of course Granjon units. Mr. Rogers's ornaments are now owned by Yale.

And finally, figure 27 was issued from the Press of Rosemary and Elliot Offner, the Rosemary Press, which uses a great deal of historic ornament.

3 REFLECTIONS

SIR JOHN DAVIES 1569–1626   **Excerpts from Orchestra or, a Poem on Dancing**

Dancing, bright lady, then began to be
When the first seeds whereof the world did spring,
The fire air earth and water, did agree
By Love's persuasion, nature's mighty king,
To leave their first discorded combating
And in a dance such measure to observe
As all the world their motion should preserve.

Since when they still are carried in a round,
And changing come one in another's place;
Yet do they neither mingle nor confound,
But every one doth keep the bounded space
Wherein the dance doth bid it turn or trace.
This wondrous miracle did Love devise,
For dancing is love's proper exercise.

Like this he framed the gods' eternal bower,
And of a shapeless and confused mass,
By his through-piercing and digesting power,
The turning vault of heaven formed was,
Whose starry wheels he hath so made to pass,
As that their movings do a music frame,
And they themselves still dance unto the same.

Under that spangled sky five wandering flames,
Besides the king of day and queen of night,
Are wheel'd around, all in their sundry frames,
And all in sundry measures do delight;
Yet altogether keep no measure right;
For by itself each doth itself advance,
And by itself each doth a galliard dance.

And now behold your tender nurse, the air,
And common neighbour that aye runs around;
How many pictures and impressions fair
Within her empty regions are there found

Which to your senses dancing do propound?
For what are breath, speech, echoes, music, winds,
But dancings of the air, in sundry kinds?

What makes the vine about the elm to dance
With turnings, windings, and embracements round?
What makes the lodestone to the north advance
His subtle point, as if from thence he found
His chief attractive virtue to redound?
Kind nature first doth cause all things to love;
Love makes them dance, and in just order move.

Hark how the birds do sing, and mark then how,
Jump with the modulation of their lays,
They lightly leap and skip from bough to bough;
Yet do the cranes deserve a greater praise,
Which keep such measure in their airy ways
As when they all in order ranked are
They make a perfect form triangular.

Learn then to dance, you that are princes born,
And lawful lords of earthly creatures all;
Imitate them, and thereof take no scorn,
For this new art to them is natural.
And imitate the stars celestial;
For when pale Death your vital twist shall sever,
Your better parts must dance with them forever.

GEORGE FLECK

Symmetry: Making the Unseeable Imaginable

One of the remarkable chemical achievements of the past century has been the elucidation of molecular structures. Fantastically small, with dynamic structures precariously maintained by the balancing of a myriad of electrical charges, molecules have the forms and the functions that give rise, in large aggregates, to all the variousness of the tangible world of matter.

Molecules can be imagined to be constructed of charged particles called electrons and nuclei, but in the molecular assemblage the electrons lose their identities as discrete particles. The electrons probably constitute a molecular cloud of negative electricity that envelopes the nuclei. This cloud is dense near the nuclei and also between certain pairs of nuclei that are said to be chemically bonded together. The cloud is diffuse around the outside of the molecule. Just as the atmosphere of the planet earth has no clearly defined outer boundary, so the electronic atmosphere of a molecule has no outer surface.

Precise and reliable experimental methods have been developed to ascertain the relative positions of nuclei in a particular molecule, and to map the density of the electron cloud throughout the molecule. Among these powerful instrumental methods are techniques involving the *diffraction* of X-rays, electron beams, and neutron beams as they pass through a substance; the *absorption* of radiation (microwave, infra-red, visible, or ultraviolet) by a substance; and the interaction of the substance with magnetic fields (nuclear magnetic resonance spectroscopy is an important example). Quantitative interpretation of such experimental data to yield structural information about molecules often involves rather sophisticated use of symmetry theory. There has been intense research activity in devising ways to use the principles of quantum theory to predict quantitatively these same nuclear positions and electron-density maps. Skillful use of symmetry theory is required for making most of these predictions. Symmetry considerations can often reduce the complexity of a molecular quantum mechanics problem so that an initially unmangeably complex calculation can be tackled. Such continuing efforts by chemical theoreticians make use of rather sophisticated specialized mathematics and large computers.

The agreement between *ab initio* theoretical calculations and experiments for most molecules does not satisfy most chemists today, although the agreement is remarkably good compared with the situation a generation ago. There also continues to be a great deal of creative effort by chemists to devise means to present the quantitative data from experiments, as well as the quantitative predictions of theory, in ways that convey vividly and faithfully the significant features of the shape and structure of a molecule. Long lists of numbers from a computer simply won't do.

Every chemical transformation, every chemical reaction, involves a reshaping of molecular electronic atmospheres. Thus it is that chemists who want to talk about molecular mechanisms of chemical reactions need to talk about the shapes of molecules and in particular about the shapes of their surrounding electron clouds. That talking is not easy to do effectively. If a molecule fades off in all directions forever, with no outer boundary, how can anyone speak of the size and shape of that molecule with any hope of precise communication? How can pencil and paper, or chalk and blackboard, be used to represent the spatial features of a molecular electronic atmosphere? We have here a significant communication problem for chemists. This problem in communication occurs, in fact, with respect to almost all the properties of molecules. Familiar words needed for properly describing a molecule do not seem to be in our everyday language. The problem arises both from the nature of molecules and from the nature of our language.

After all, our language was developed in large part to deal with a world that is directly experienced and perceived by the senses. This is the world of objects and materials that we can touch, see, taste, smell, and hear. This is the world that can be photographed and recorded, that can be shared easily with others, often simply by two persons being in the same place and sensitive to the same sights and sounds. Often this concrete tangible world appears to be our real world. Some say that it is the only real world.

To be sure, people think and talk about other realms, and imaginations seem unbounded and unconstrained as art, literature, and music speak of worlds not directly accessible to the senses. These languages of the imagination are necessarily abstract, are sometimes vague and ambiguous, and often convey different impressions to different people. These are not always worlds that can be easily shared with others.

For a chemist the world of molecules is in part a world of the imagination. Molecules cannot be directly perceived as individuals. Molecules don't have color, texture, strength, hardness, flavor, shape, or size—not, at least, in the ordinary sense of those words. Words linked closely to our perceptions of things like a baseball, a plucked guitar string, clouds in the sky, or wine in a glass are apt to be misleading when used to describe molecules. Abstract and fanciful language, with a minimum of reference to the tangible world, quite often seems best suited for talk about molecules. Sometimes a theoretician may choose to describe some aspect of a molecule by a long, computer-generated list of numbers. Or indeed a property of a molecule may be expressed as a collection of mathematical equations that no one knows how to evaluate.

Yet there really are chemical substances, and there really are molecules. If indeed all the tangible world owes its existence to the properties of molecules, then the world of molecules is real, even if it is substantively realized only in the chemist's imagination. Any reasonable description of a molecule must be consistent with the results of experimental investigations. A chemist's imagination should operate within the confines of constraints that come from experimental data and from trustworthy theories, but that imagination should not be restrained by problems essentially linguistic in nature. And the chemist must be able to communicate with other scientists—and with lay audiences—about molecules.

A significant challenge for contemporary chemists is the development of useful and vivid descriptions of molecules, descriptions that are faithful to the spirit of physical theory and consistent with experimental observations, but which are also sufficiently abstract to permit both speaker and listener to make creative use of their imaginations. The goal is a rich and vivid mind's picture that is not demonstrably false in any respect, a description that frees a speculative person to think constructively about the properties and interactions of specific molecules.

Models for Molecules

As early as 1810, John Dalton used wooden models made from balls and sticks to illustrate the combination of atoms to form compounds. He found these models to be effective teaching tools. A. W. Hofmann used a collection of elaborate croquet-ball models of molecules to illustrate his Royal Institution of Great Britain lectures in 1865 (fig. 1). Others of a later generation called their smaller models tinker-toys. Almost every chemist now keeps such a set of molecular models close to his desk, finding them convenient and sometimes necessary devices to aid in visualizing particular molecules. There are many varieties now available, but wooden balls with holes drilled in them, and sticks to fit into the holes, serve well. The balls are nuclei; the sticks are bonds.

Ball-and-stick models can tell us many things about the way in which

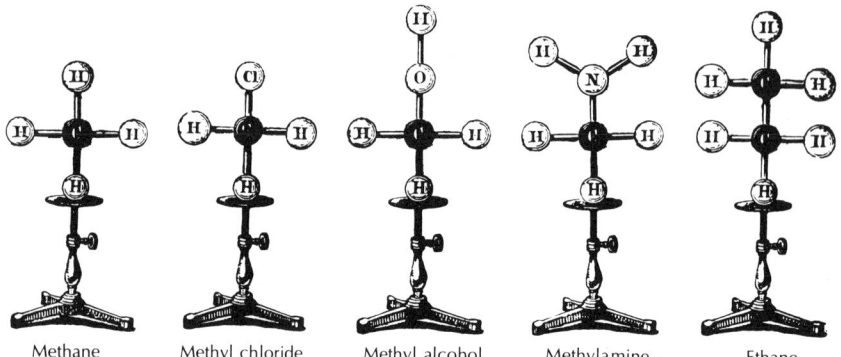

Methane   Methyl chloride   Methyl alcohol   Methylamine   Ethane

1. Models of molecules, constructed from croquet balls, used to illustrate a lecture in 1865. A twentieth-century chemist would construct the models so as to achieve a tetrahedral arrangement of balls around each carbon atom. After Hofmann, A. W., *Proceedings of the Royal Institution of Great Britain*, vol. 4 (1865), p. 416.

nuclei are arranged in space. But the models may imply more to the viewer than the model builder intends. If the balls are meant to represent nuclei, then these balls grossly exaggerate the size of the nuclei relative to the overall size of the molecule. If the balls are meant to represent some portion of the electron cloud, then we have the problems of representing an unbounded and resilient cloud by a hard-surfaced ball. The chemical bonds in a real molecule surely cannot be much like the sticks that hold the wooden or plastic balls together, yet the sticks are located in space where bonds might be, and the sticks are the physical connections between the balls, just as the bonds hold together the nuclei in a molecule. A well-built model does not fall apart easily, yet this very stability belies the essential dynamic aspect of any molecule. Many models have color-coded parts, with perhaps black balls for carbon, red balls for oxygen, and so forth. These colorful models are excellent for many purposes, but they subtly suggest that carbon and oxygen "look" different in the actual molecule. Models built of wood, plastic, or metal are intrinsically misleading. Only the very wary model user can keep from associating some of the material features of his model with his mental image of the molecule itself. Every model tells, at the same time, not enough of the facts and too much of the fiction about the invisible, untouchable molecule.

Ball-and-stick models of molecules sometimes have shapes that suggest analogies involving everyday objects such as envelopes, chairs, boats, propellers, sandwiches, and such. Extensions of these analogies can give rise to an improbable jargon that mystifies outsiders, even other chemists. Consider, for example, the vocabulary often used for describing two conformations of a molecule of cyclohexane, C_6H_{12}. A ball-and-stick model of cyclohexane can be constructed by fitting together a ring of six carbon "atoms" connected with six "bonds." The holes in the balls that represent carbon atoms are drilled with four tetrahedrally oriented holes for the sticks. The resulting carbon skeleton can be twisted into two conformations: the puckered *boat* conformation shown in figure 2a, and the *chair* conformation shown in figure 2b. When sticks are placed in all vacant holes (and hydrogen-atom balls placed on the ends of all these sticks), one finds that there are two kinds of carbon-hydrogen bonds, differing in orientation: these are called *axial* and *equatorial* bonds. The adjectives are only loosely related to the ordinary meanings of axis and equator. In a model of a boat-form cyclohexane molecule, there are four types of carbon-hydrogen bonds: the two types at the sides of the boat are called boat-equatorial and boat-axial, and those at the ends of the boat are called flagpole and bowsprit (fig. 2c). Chemists then speak of flagpole-flagpole interactions between hydrogens. It is important that the novice have a clear understanding about just how many of the qualities of two interacting flagpoles are being ascribed to the cyclohexane molecule.

There is one significant property that is shared by molecule and proper model: the symmetry properties of the model are precisely the same as the symmetry properties of the molecule. This fact serves as a useful guide for a chemist's imagination. If we take a collection of proper models of various sorts, each representing the same molecule, then we can ask about geometric properties shared by all. The molecule itself may also

2. Ball-and-stick models (on the right) of two conformations of cyclohexane. The balls represent carbon atoms. The spatial arrangement is perhaps more easily visualized in the drawings on the left, in which each straight line connects two carbon atoms. (a) This model possesses two mutually perpendicular mirror planes; the intersection of the mirrors is a two-fold axis of rotation. Some chemists call this the *boat* conformation. (b) This model contains three mirror planes, and it also has a center of inversion. It is sometimes called the *chair* conformation. (c) In cyclohexane, each carbon atom is connected to two hydrogen atoms, and here four sticks are in place to show that there are four different spatial orientations for these hydrogen atoms. Since this is a boat, what could be clearer than to label positions with names such as bowsprit and flagpole? The complete stick model has more flagpoles, bowsprits, and so forth.

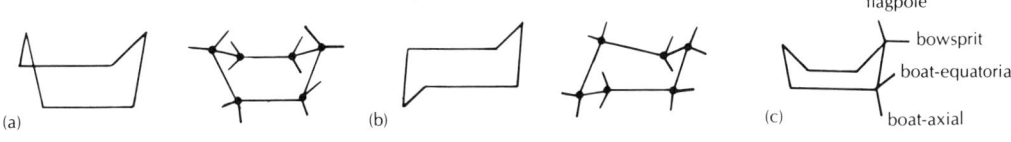

share these properties; it does share the symmetry properties.

Other models exist that are not subject to the limitations of wood, plastic, and metal—models that can be made to tell a lot of the truth and very little fiction. These are the numerical models that arise from approximate quantum mechanical calculations. Often such a model does not dwell on the nuclei, but instead focuses on the structure of the enveloping electron cloud. One such description presents a collection of numbers, representing the electron density at selected points throughout the molecule. These numbers can be presented graphically by means of electron-density contour maps drawn on sections cut through the molecule. The collection of numbers possesses the same symmetry properties as does the molecule it represents.

The guiding symmetry principle, useful both in making the approximate calculations and interpreting the resulting numbers, is that *the symmetry of the collection of nuclei is the same as the symmetry of the electron density distribution, and each of those symmetries is the same as the molecule as a whole.* This principle, valid for any isolated molecule, is a consequence of the dynamic nature of each molecule and of the forces that maintain its integrity. The electrons are free to move wherever they will, subject only to electrostatic forces (attractions from the nuclei, repulsions from each other) and the constraints of quantization of energy and of the Pauli exclusion principle. The forces attracting the electrons necessarily have the symmetry of the collection of nuclei; in the absence of external forces, the electrons must assume a distribution of precisely the same symmetry. The same considerations also apply to the whole aggregate of reactants during the course of a chemical reaction, as the electrons rearrange to break bonds and form bonds while always maintaining the symmetry of the instantaneous nuclear configuration of the interacting molecules. The fundamental requirement is that the collection of positively charged particles (the nuclei) and the collection of negatively charged particles (the electrons) mutually determine each other's distribution in space. Neither collection can have an element of symmetry missing in the other.

Symmetry as a Guide to Responsible Chemical Imaginings

It is convenient for chemists (and in the absence of any evidence to the contrary, it seems safe) to assume that properties of three-dimensional space remain invariant as one moves from the familiar macroscopic world to the invisible molecular world. In molecules the concept of a localized particle becomes unreliable, the possibility of continuously variable energy disappears, but we assume that the nature of space stays the same. This assumption permits symmetry principles and geometric principles, developed and refined in our familiar world, to be used in describing and in imagining a molecule. We use the principles of symmetry to calibrate our intuition about the world of molecular dimensions.

It is important to have some reliable guideposts when studying molecules, because reputable authorities put forth ideas that appear very different, and because both the real progress in molecular chemistry and the changing whims of fashion introduce rapid change in the research literature and in textbook descriptions of molecules and molecular processes. There are two important factors that need to be understood and placed into perspective by students of chemistry:

(1) Since the beginnings of chemistry, and continuing to the present time, many chemists have endowed their favorite molecules with attributes beyond those supported by experimental data. Sir Benjamin Brodie, the controversial Oxford chemist, strongly urged his colleagues during the 1860s and 1870s to use only those theories about molecules that were required by observations, proposing his nongeometrical calculus of chemical operations to replace the ball-and-stick models that were becoming increasingly popular. Indeed, the overwhelming majority of physical chemists in the late nineteenth century explicitly rejected geometrical ideas about molecules (and many rejected the idea of molecules as well) as being neither required by experimental data nor needed for research in chemistry. Nevertheless, this same period of time saw a vast outpouring of research literature dealing with the structural features of molecules. These chemists usually went well beyond the requirements of their data in proposing molecular structures, allowing their imaginations substantial latitude. The historical record shows that the procedure of thinking about molecules in terms of rather explicit models, restrained only by the requirement that nothing in the models contradict experiment, resulted in an extremely productive period of chemical research during which the field of structural organic chemistry, and soon thereafter the field of structural inorganic chemistry, developed and flourished. The strategy is still being used, and still with success. But today, now that sophisticated experimental methods are yielding a wealth of information about the details of molecular structure, a note of caution is appropriate: long acquaintance with a molecule tends to blur the distinction between the properties required by experiment, and the properties bestowed by kindly speculators. Even when respected authorities are speaking about "their" molecules, the listener should be skeptical when the descriptions become vivid, the attributes realistic.

(2) Detailed theories of electrons in molecules have changed during the past decades at a revolutionary pace. This evolution and upheaval continues. The probability is high that many details of the predictions of approximate quantum mechanical theories relating to a particular molecule will change several times during the professional lifetime of a now twenty-year-old chemist.

These two factors—the psychological tendency of many good chemists to flesh out their conceptions of molecules beyond the strict requirements of experiments, and the rapid evolution of the theories of molecular quantum mechanics—combine to make needed some ways of sifting and winnowing the grains of reliable information from the chaff that the winds of change may soon blow away.

The conclusions founded on symmetry theory can be safe and sure. While such descriptions of molecules may be sparse, they are closely related to experimental results and to apparently safe extrapolations about the nature of three-dimensional space. The language of such conclusions is often that of mathematical group theory. Fundamentally, the logic involves an analysis of how one can fit things into space, of how one can fill space with things. There is an intimate relationship between the conclusions and the methodology of the applications of symmetry to the study of whole crystals, and to the study of individual molecules. And there is also a very real kinship with the many other applications of group theory and of symmetry principles, including formal relationships between the study of molecular structure and such apparently diverse fields as bell ringing, weaving of textiles, the dance, many aspects of the graphic arts, and many forms of music.

Chemists have sometimes spoken vividly of their molecules, but the colorful language has gone beyond the information available from the laboratory to imply molecular qualities that may not exist. Descriptions tied closely to the strict requirements of experimental data have often been drab, uninformative, and not particularly stimulating to the imagination. It might well be a fruitful search for some young chemists to look for a more vivid and yet precise method of communicating about molecules that draws upon analogies from other areas of man's artistic and creative activities where the principles of symmetry and group theory have important applications. For instance, the precision of carefully stated symmetry properties combined with the study of dances having the same symmetries might be illuminating in thinking about molecular vibrations.

With symmetry as a guide, the unseen and unseeable molecules can be imagined, and our imaginings can be faithful to all that the laboratories have been able to tell us. And that is something, indeed.

ALLAN LUDMAN

Symmetry: The Framework of the Earth

To a casual observer, symmetry would seem to have negligible application to geology. Certainly there is little symmetry in the violent eruption of a volcano, the devastating release of energy in great earthquakes, or the distribution of continents and ocean basins on the surface of our planet. In fact, asymmetry might appear to be the rule. Erosional and depositional phenomena associated with streams, glacial ice, and wind are commonly asymmetrical, and the very asymmetry of such features as sand dunes, current ripplemarks, drumlins, and roches moutoneé is of value to the geologist in interpreting events of the far-distant past. Why is it then, that a significant part of the training of every geologist is a thorough exposure to the symmetry operations and to the detection of symmetry elements?

The answers to this question permeate nearly every aspect of geology, from the submicroscopic realms of atomic structures of minerals to the megascopic realm of drifting plates of earth material the size of continents. Some students are surprised to find, after several years of courses in the many disciplines of geology, that symmetry principles are the means of understanding processes from diverse disciplines and integrating them into a comprehensive model for earth dynamism. Some students never realize this, and a valuable synthesizing concept is lost to them.

The beginings of the science of geology lie in the entirely pragmatic search by early man for materials which could enhance his chances for survival. A brief survey of the development of the discipline of mineralogy may prove helpful in demonstrating how an approach based on symmetry has been utilized for over 300 years in order to unlock the mysteries of the mineral kingdom. Techniques have changed as sister sciences provided new instrumentation, but the thread of symmetry has continued unbroken to the present day. In 1669, after years of meticulous measurements of thousands of crystals, Nils Stenson (Nicolaus Steno) noted that the angular relationship between a pair of crystal faces on one specimen of a mineral is identical to that between the same pair of faces on every other specimen of that mineral, regardless of size of specimen or degree of perfection achieved during growth. This implied that there must be some sort of regular internal structure associated with individual mineral species, and that careful examination of external mineral morphology could lead to an understanding of the (at that time) invisible and impenetrable interior of the mineral. Such studies occupied mineralogists for the next two centuries.

The external form of a mineral species may vary. Garnet, for example, occurs in crystals of 12 faces (rhombic dodecahedron), 24 faces (trapezohedron), or 36 faces (a combination of the two). But, the combination of symmetry elements for *every* crystal of garnet—and for every crystal in a mineral species—is always the same. Measurements of interfacial angles of minerals by the optical goniometer showed that minerals could be classified into a small number of groups based on their combinations of symmetry elements. A single *crystal class* contained all minerals possessing the same symmetry formula—same number of axes of rotation and rotary inversion, same number of planes of symmetry, and same relationship between planes and axes. In such a classification, such dissimilar minerals as galena (PbS; metallic, gray, high specific gravity), halite (NaCl: nonmetallic, colorless, low specific gravity), and diamond are grouped with garnet even though these minerals rarely display dodecahedron or trapezohedron faces. Despite the fact that there are nearly 3,000 minerals known, there are only 32 crystal classes.

Since minerals with strikingly different physical properties, chemical compositions, and modes of genesis are grouped together by symmetry of the crystal form, the inference from Steno's work was reinforced: similarities in external form must be caused by similarity in internal structure. Long before atomic theory was applied to the mineral kingdom, Haüy suggested that every mineral is constructed by repetition in three dimensions of a relatively simple building block whose symmetry is that of the crystal class to which the mineral belongs. In 1912 the "impenetrable" interior of solids was pierced for the first time when Laue utilized X-radiation to demonstrate the existence of regular atomic patterns in crystals. With further X-ray studies, it became apparent, as the early crystallographers predicted, that the external symmetry of minerals is caused by a symmetrical arrangement of ions in the mineral lattice. Soon, evidence of symmetry operations other than the classic rotation and reflection were discovered. In a three-dimensional array of ions, the operation of translation combined with rotation and reflection results in screw axes and glide planes respectively.

The unifying, interdisciplinary nature of symmetry was clearly demonstrated by these discoveries and by parallel developments in mathematics and chemistry. Students of mathematical group theory had proved the existence of 32 crystallographic *point groups,* each defined by a specific combination of symmetry elements. These are the crystal classes of the mineralogist. Mathematicians found 230 abstract *space groups.* Concrete examples of many are found in the structures of minerals as determined by X-ray studies. Using group theory to study electron density probabilities in molecules, chemists were able to predict what the structure of compounds should be. Tangible evidence for the correctness of their deductions comes from the rigid frameworks of the crystalline compounds called minerals.

On a mesoscopic scale, symmetry is also part of the language and technique of the sedimentologist, the paleontologist, and the structural geologist. Modern ripplemarks in sand or ancient ones in sandstones may be symmetrical or asymmetrical. The difference is caused by basic differences in the flow of the fluid responsible for producing the ripple form, so that recognition of symmetry (or lack of it) in such a feature can be helpful in reconstructing the ancient environment in which the rock formed. Students of elementary paleontology learn to distinguish bivalves from brachiopods by the orientation of the single plane of symmetry possessed by the organism. A structural geologist studying a sequence of folded strata determines whether a given fold is symmetrical or asymmetrical, and whether the fold style in a map area, as revealed by detailed analysis of many folds, possesses orthorhombic or monoclinic symmetry. The geomorphologist studying glacial phenomena, the geophysicist probing the internal structure of the earth with seismic waves, the optical mineralogist, the ceramist—all these scientists must understand and work with the basic principles of symmetry on a day-to-day basis.

Because symmetry principles are so pervasive in the fabric of geology, it is only fitting that one of the most exciting and significant advances in the science has come about because of the recognition of a plane of symmetry where none had ever been suspected. Since the beginnings of modern geology, debates have raged over the causes of mountain building, the reasons for the present distribution of continents and oceans, the age and permanence of the oceans. Modern oceanographic surveys revealed the presence of a ridge or ridge system in every ocean, frequently associated with volcanic activity, as at Iceland. Geophysical studies of the ocean basins detected systematic anomalies in the magnetic fields of the ocean basins, the anomalies generally oriented parallel to the oceanic ridges.

The landmark discovery was that the oceanic ridge is a plane of symmetry. The anomalies on one side are mirror images of those on the other. Further, ages of extinct volcanoes on opposite sides of the ridge increase symmetrically with distance from the ridge. It is now known that the ages of the anomalies also increase outward from the ridge. Modern plate tectonic theory holds that the ridges are zones along which the earth's crust is spreading apart, zones which act like dual conveyor belts carrying crustal material away from the ridge crests. With these discoveries, the concept of continental drift left the doghouse of geologic ridicule and became the vehicle for geologic research in the 1970s. A plane of symmetry thus led the way, but on such a scale as was never dreamed of.

Applications of symmetry are often hidden from the eyes of the untrained observer. Hidden therefore is the relationship through symmetry of geology, chemistry, biology, mathematics, literature, art, music, architecture, dance, and many, many more aspects of our world. Symmetry is the common thread which unites art and science, artist and scientist. Its applications are endless.

MARJORIE SENECHAL

Symmetry: The Perception of Order

I. Form

Once we become aware of symmetry, we suddenly find it everywhere we look: animals, vegetables, minerals; textiles, floors, and walls. In music, too, there is symmetry; the repetition is in time instead of space. Yet, if we look closely, we see that none of the symmetry which surrounds us is exact. Snowflakes are irregular, tiles are uneven, starfish do not really have five equal arms. The symmetry we seem to see exists only in our mind's eye; our real eyes see only departures from it.

The symmetrical Renaissance dances performed at the Symmetry Festival by the Cambridge Court Dancers reflected the belief of that time in an ordered and harmonious universe, and in the divine ordering of society according to the same plan. The sixteenth-century English poet Sir John Davies, in his long poem "Orchestra," celebrated this highly patterned dancing by citing instance after instance of analogous dances in nature. Even in those days there were dissenting ideas, but Sir John dismissed them:

> Or if this all, which round about we see,
> As idle Morpheus some sick brains hath taught,
> Of individual motes compacted be,
> How was this goodly architecture wrought?
> Or by what means were they together brought?
> They err that say they did concur by chance:
> Love made them meet in a well-ordered dance!

Today, four hundred years later, our belief in a unified natural and social order has all but vanished. The Uncertainty Principle has become a fundamental tenet of science; chaos seems to reign in the universe. The purpose of education is often said to be preparation for coping with a world of uncertainty and change.

And yet we celebrated a Symmetry Festival. But the purpose of the festival was not to evoke nostalgia for a lost vision of universal constancy; rather, we were celebrating our ability to distinguish between what is constant and what is not. The human mind can discern regularity of form beneath any obscuring variation. And although everything departs from its abstract ideal, the ideal remains of value to us because it provides a norm against which we can measure the reality we find in the natural world.

II. Patterns

Mathematics is the study of ideal forms, often abstracted from natural ones, and in the experimental sciences many of these forms are tested as frameworks for the clarification of complicated phenomena. The geometrical theory of symmetry and its applications illustrate this symbiotic relationship between mathematics and experimental science. This theory explains the geometry of regular packing arrangements (repeating patterns) and the algebraic analysis of the relationships upon which the repetition is based. The ideal forms studied have been abstracted from decorative ornaments and from the striking geometric forms of crystals. The resulting theory has in turn provided the systematization needed for the understanding of internal crystal structure.

The theory of patterns has its origins in the most ancient mathematics. The ingenuity exhibited in ancient ornamental art leads us to conclude that the artists were conscious of the abstract problem of paving and tiling the plane. As Hermann Weyl points out in his book, *Symmetry* (Princeton, 1952):

> Examples for all 17 groups of symmetry are found among the decorative patterns of antiquity, in particular among the Egyptian ornaments. . . . The art of ornament contains in implicit form the oldest piece of higher mathematics known to us.

This early mathematics must have arisen from the perception of systems of symmetry in the earliest and most simple decorative patterns. Another early source for the development of a theory of symmetry was probably crystal form. The ancient Greeks discovered the five regular solids (fig. 1) and considered them to be the building blocks of the material world: the cube was associated with earth; the octahedron with air; the tetrahedron with fire, the icosahedron with water; and the dodecahedron with the cosmos. The first three of these solids are common crystal forms, but the last two are never found in nature. Perhaps the latter were discovered through mathematical investigation, extrapolation of the symmetry properties of the first three, or perhaps the form of the dodecahedron was inspired by the mineral pyrite, an irregular pentagonal dodecahedron. The icosahedron is related to the dodecahedron in the same way the octahedron is related to the cube (each can be produced by connecting the centers of adjacent faces by straight lines), and so its construction would have been a logical sequel.

The theories of the symmetry of plane patterns and finite forms thus undoubtedly had their origins in the visible world. That a hidden but

93 Symmetry: The Perception of Order

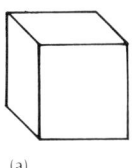 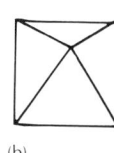 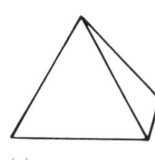 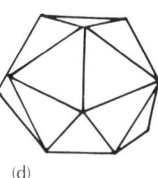 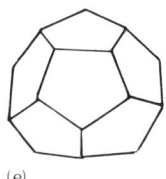

(a)   (b)   (c)   (d)   (e)

1. The five regular solids were known to the ancient Greeks. (a) cube, (b) octahedron, (c) tetrahedron, (d) icosahedron, (e) dodecahedron.

profound relationship might exist between them was surmised following the study of cleavage in crystals by Haüy about 1800. After accidentally dropping and breaking a crystal of calcite, he observed that the fragments seemed to be miniatures of the original. The fragments themselves could be broken into smaller and smaller copies. Haüy suggested that crystals were ultimately constructed of submicroscopic building blocks. By choosing suitably shaped blocks and appropriate ways of stacking them, one could reconstruct the external forms of perfect crystals. If this hypothesis were correct, then the shape and arrangement of the building blocks would provide clues to the ideal shape of the crystals, and conversely the shape of the crystal would provide information about the shapes of the building blocks. This hypothesis gave impetus to a further development in pattern theory: the extension of the two-dimensional theory to the packing of objects in three-dimensional space.

The theory of two- and three-dimensional patterns led to significant conclusions about form in nature. For example, it explained the nonexistence of certain crystal forms. The kinds of symmetries possible in repeating patterns are limited to the symmetries of the shapes that can be used to build them: they are constrained by the ways in which polygons can be fitted together to fill a plane surface, or polyhedra to fill space. If you look closely, you will find that wallpaper and tile patterns are always based on rectangles, triangles, squares, or hexagons; such patterns have the two-, three-, four-, or six-fold rotational symmetries of their polygonal units (fig. 2). Five-fold, seven-fold, or higher-fold rotations are impossible in plane patterns. The reason is that these rotational symmetries are incompatible with packing requirements. For example, a plane surface cannot be completely covered with regular pentagons (fig. 3a). Or, to state this conclusion algebraically, if five-fold rotation is a symmetry operation for a figure, then translation is not; conversely, if a design has translational symmetry (if it is an infinite repeating pattern), then it cannot have axes of

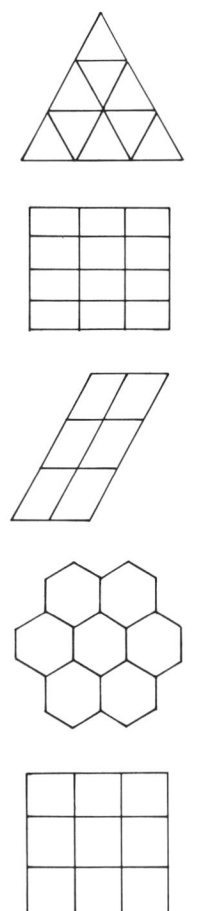

2. Every plane pattern is based on a network of parallelograms, rectangles, triangles, squares, or hexagons.

five-fold rotation. Thus, though pyrite specimens are found in abundance in many parts of the world (fig. 3b), no prospector will ever find a perfect pentagonal dodecahedron among them—not because they are hard to find, but because we can prove that this symmetry is impossible for a crystal. If a plane pattern cannot have pentagonal symmetry, neither can a stack of plane patterns; consequently, no form constructed of layers of repeating patterns can have it either. If the hypothesis that the atoms of a crystal are arranged in pattern layers were true, no crystal could assume the shape of a regular dodecahedron or icosahedron.

Until 1912, the theory of patterns remained only a hypothetical model for the internal structure of crystals; no one could be sure whether the model was correct, because the arrangements of atoms could not be seen. In 1912 the discovery of the diffraction of X-rays by crystals proved indirectly that the atoms do indeed form regular three-dimensional patterns, and geometric pattern theory was immediately employed in the interpretation of the diffraction photographs and classification of crystals. More recently, the theory of patterns has begun to influence the course of research in the structure of living organisms. After many successes in determining the structures of mineral and organic crystals, attention also turned toward the diffraction patterns of virus crystals. Viruses are the simplest of self-reproducing organisms. They lie on the borderline of life, more complex than any nonliving chemical complex but simpler than any living thing. A virus crystal is a crystal built of viruses, just as organic and inorganic crystals are built of molecules, atoms, or ions. Crystallographers hoped that a study of the symmetries of the virus crystals would give clues to the structures of the individual viruses.

X-ray photographs soon showed that virus crystals have a high order of symmetry, and from the theory of patterns we conclude that individual virus particles must have such symmetry also. In an effort to relate this general conclusion to the fine structure of virus particles, in 1956 J. Watson and F. Crick published a short article in *Nature* on the "Structure of Small Viruses." (Although better known for their model of DNA, Watson and Crick also gave direction to a whole field of scientific activity when they published the virus study.) They knew that viruses are composed of molecules of protein and RNA; the latter is the infective agent, and the former acts as a protection for it. They wanted to find how a virus is put together. From the experimental data then available and the hypothesis that "wherever, on the molecular level, a structure of a definite size and shape has to be built up from smaller units ... the packing arrangements are likely to be repeated again and again and hence the subunits are likely to be related by symmetry elements," Watson and Crick concluded that the symmetry group of a virus which appears to be a polyhedron must be that of one of the five regular solids. Their suggestion led to the careful investigation of virus shapes and to the discovery that most small polyhedral viruses have the symmetry of the icosahedron.

This says something very important about the structure of these viruses: they are certainly not constructed of parallel layers. Pentagonal symmetry can exist if the units are packed around a sphere, but not if they are stacked. Thus a knowledge of the symmetry of viruses and the role of that symmetry in the abstract theory tells us clearly that polyhedral viruses are not built of layers like crystals but instead probably consist of RNA inside a spherical protein shell. This does not mean that the exact structure of the viruses is immediately known, but it does mean that from the myriad of conceivable forms, we can clearly separate those that are possible models from those that are not.

And the story does not end here; the problem of determining virus structure raises many questions—some old but unsolved, some new—about noncrystallographic packing arrangements in space.

III. Design

". . . the packing arrangements are likely to be used again and again" . . . on the molecular level, and on other levels too: in crystals, wallpaper, and architecture, in music, dance and art. Perhaps it is this that gives rise to "this all, which round about we see." The rhythms of the "well-ordered dance" receive their beats from symmetry.

The rhythms are not perfect; there are inherent variations and accidental ones. We can never predict how closely a particular growing crystal or plant will eventually resemble its ideal form, but we expect that on the average, most will do so rather more than less; and the crystal more than the plant, because of the far greater complexity of living systems. This statistical interpretation of symmetry incorporates our awareness of the role of chance. Symmetry becomes the norm to which all things approximate; it is the fundamental blueprint which nothing is expected to follow to the letter.

The theory of symmetry is a triumph of the human intellect. It is the perception of order in a chaotic universe, the study of the forms that order can take, and the use of that study to give significance to the things we see. In science as well as in the arts, symmetry is the geometric plan on which the variations of nature and of life are drawn.

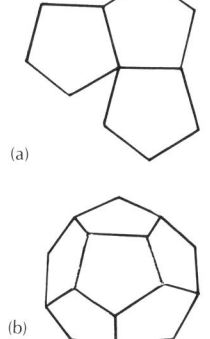

3. (a) Regular pentagons cannot be fitted together to pave a plane surface. (b) The pentagonal faces of dodecahedral pyrite are irregular because no crystal can have five-fold symmetry.

PHILIP REID

Symmetries in Plants

The study of plant growth and development, which has been labeled morphogenesis, is the most all-encompassing, the most difficult, and the least understood aspect of plant biology. This is so because it involves changes in form. The ultimate expression of these changes illustrates interesting and aesthetically pleasing symmetries on several different levels. The most obvious is the overall symmetry of the plant which gives the distinctive shape to each species and allows us to recognize the shape of an elm tree, for example. The external symmetry of the individual shoots and roots of the organism is somewhat less obvious, as is the symmetry of the individual parts such as leaves and flowers. And lastly, the symmetry of the internal structures is visible only with the aid of a microscope.

The difficulty in understanding the changes which take place at both shoot and root apices lies in the fact that growth is influenced by a myriad of both endogenous and exogenous factors. While many of these factors (such as hormones and light) have been isolated and identified, their modes of action for the most part remain a matter of speculation. Not only is the expressed growth pattern the result of all of these factors acting simultaneously, but it is also often the result of the complicated interaction of these factors. Furthermore, these factors may act differently in different species. Discussions of symmetry in the plant kingdom must therefore remain descriptive until the details of growth and developmental control can be elucidated.

While all plant groups show interesting symmetries, the study of growth and development in angiosperms (plants which produce flowers) is of particular interest since it is this group with which the nonbotanist is most familiar. This discussion will only deal with symmetry as it is observed and as it develops as growth proceeds in certain angiosperms.

The basic plant axis consists of a shoot and a root, and growth of the plant is the result of the production of new cells and their subsequent elongation which occurs near the apices of these two organs. The symmetry which we observe along this axis is caused by the way in which lateral appendages are distributed. The lateral appendages of the shoot are leaves and buds, the latter representing embryonic shoot systems which can ultimately produce new shoots, usually with symmetry exactly like that of the stem on which they occur. New leaves and buds are initiated at

97 Symmetries in Plants

the apex of a stem in a region called a meristem. As one descends a stem from the apex or meristem, the lateral appendages encountered are of increasing age. Within the meristem, a dome-like structure of young rapidly dividing cells, the symmetry of the plant as a whole is determined.

Figure 1 shows a scanning electron micrograph of a shoot apex and two young leaves visible here as folds of tissue which, after initiation, differentiate into structures recognizable with the naked eye as leaves. It is obvious, then, that the position of leaves along the stem, called phyllotaxy, is determined by the events which occur in the relatively few cells which make up the apical meristem. The phyllotactic patterns which result from these precisely regulated events are either whorled (three or more leaves at a node), opposite (two leaves at a node), or helical (one leaf at a node but initiated around the dome such that a screw axis is observed in the mature stem). It is the rotation combined with translation at the point of appendage initiation which makes the study of phyllotaxy intriguing and also difficult.

While the helical arrangement of leaves along a stem is the usual case in angiosperms, many plants (such as Coleus) show opposite phyllotaxy.

1. Scanning electron micrograph of a shoot apex of *Zea mays* (corn). Dome-like structure in the center is the apical meristem. The youngest leaf is the fold of tissue at the bottom. Next in age is the fold of tissue at the top which nearly encloses the apex. Note that in this plant new leaves are initiated opposite one another on the apical dome, but only one leaf occurs at each node. Elongation will result in helical phyllotaxy. (Photo courtesy of Otto L. Stein)

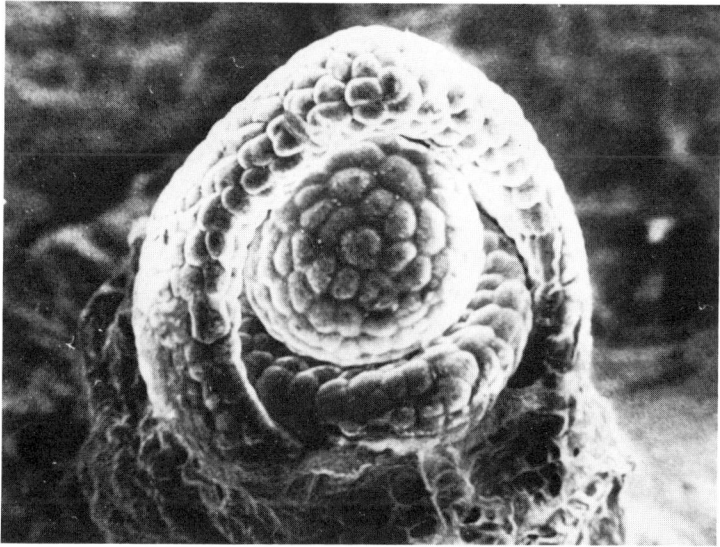

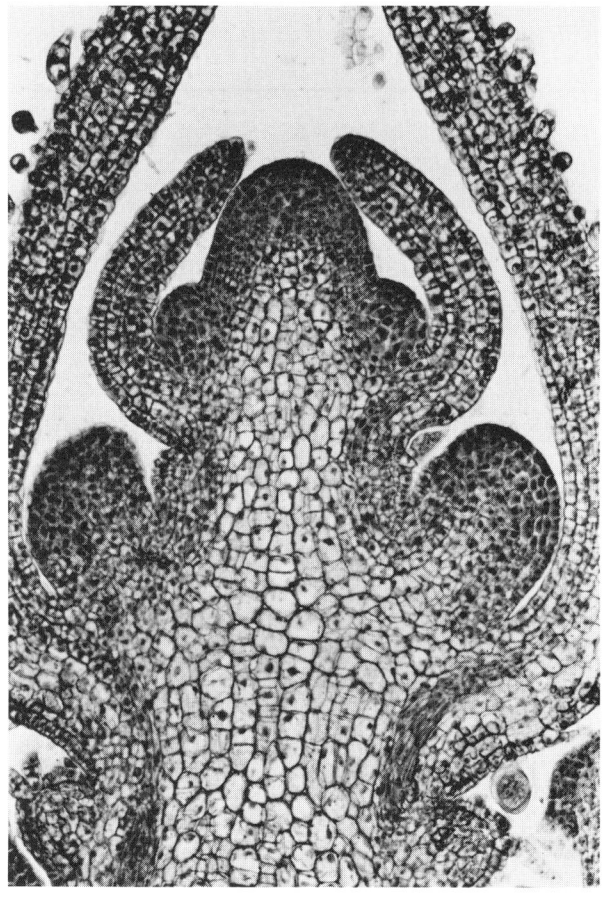

2. (Left) Longitudinal section of Coleus stem. Two leaves are initiated opposite one another on the apical dome at each node. Elongation will result in opposite phyllotaxy with each leaf pair at right angles to the pair above and below. Leaves normal to the plane of the section are removed in this preparation. 3. (Right) Sunflower head showing contact parastichies which intersect at approximately right angles. Strings were placed on the head to indicate one long and one short spiral.

Figure 2 shows a longitudinal section of such an axis, which is of course missing the pairs of appendages which are normal to the plane of the section. Such a phyllotactic pattern can be thought of as two separate screw axes separated by 180° around the stem.

Looking down a Coleus stem from the apex, one observes that there are four ranks of leaves with the younger leaves being superimposed on the older. Such a rank, or straight line of leaves, is called an orthostichy. One can also observe such vertical ranks of leaves in stems which show helical phyllotaxy. In such plants, however, the successive leaves in the orthostichy may be separated by several nodes. That is, to get from the youngest leaf in an orthostichy to the next youngest, it may be necessary to follow the helix through more than one gyre. It has been customary in the literature to describe this pattern as a fraction, the numerator of which is the number of gyres of the helix between the leaves in the same orthostichy and the denominator of which indicates the number of orthostichies in the system.

The simplest case of such leaf arrangement can be seen in *Ulmus americana* (American elm) in which the leaves are truly alternate, 180° apart at each node. One gyre of the helix is necessary to get from a given leaf to the next oldest in the orthostichy and this gyre involves two leaves. This case may be described by the fraction $\frac{1}{2}$. More complicated systems have been described which fall into the series $\frac{1}{2}, \frac{1}{3}, \frac{2}{5}, \frac{3}{8}, \frac{5}{13}, \frac{8}{21}, \frac{13}{34}, \frac{21}{55}, \frac{34}{89}$, etc. Since the number in each numerator and denominator is the sum of the two preceding it, the series is an example of the Fibonacci series of numbers. Obviously the fractions result from the angle of divergence between successive leaf initials on the apical dome. It is an interesting fact that the higher fractions approach the decimal 0.38197, or the angle 137°30'28"—the so called "ideal angle." Much has been made of this, particularly in the older literature, while more modern students of morphogenesis have concentrated on the cellular and molecular events at the apex which result in these phyllotactic patterns.

Some shoot apices which do not elongate as new appendages are formed result in an expanded apex with very closely packed appendages. The sunflower is an example of such an apex, and in such systems it is not possible to identify orthostichies. Figure 3 shows such an apex in which the appendages are positioned in logarithmic spirals radiating outward from the center. Again, the youngest members of the spiral are at the center.

In such systems it is possible to observe two sets of spirals which radiate in opposite directions and which interesect each other at approximately right angles. These spirals have been called contact parastichies. The two

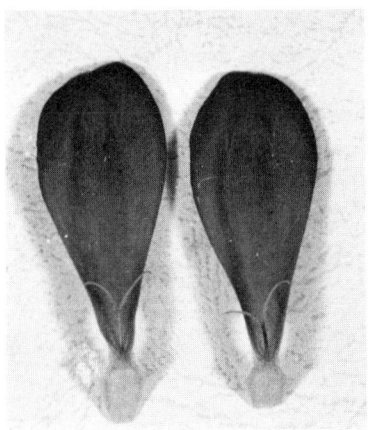

4. (Left) Individual flowers removed from a Zinnia inflorescence show dorsiventral symmetry.
5. (Right) Zinnia inflorescence. Individual flowers show dorsiventral symmetry; in the head they make a radially symmetrical pattern.

spirals are indicated here by strings, and it can be seen that one set of spirals is shorter and has fewer spirals than the other set. Again, these parastichies are often expressed as fractions with the numerator indicating the number of long parastichies and the denominator the total number in the system. Most sunflowers show 34 short parastichies and 55 long ones which radiate in the opposite direction. The fractional expression for this system would be $\frac{55}{89}$. Parastichies have also been described in other species which can again be arranged into the Fibonacci series $\frac{2}{3}, \frac{3}{5}, \frac{5}{8}, \frac{8}{13}, \frac{13}{21}, \frac{21}{34}, \frac{34}{55}, \frac{55}{89}$, etc.

The organization of root appendages is less well described, in part because root systems are more difficult to study. It should be pointed out, however, that the initiation of lateral roots takes place deep within the root tissue rather than on the surface of a dome. Most root systems show radial symmetry both externally and internally.

Most plant appendages, such as leaves and flowers, show either radial symmetry in which there is an axis of rotation around which symmetry is uniform, or dorsiventral symmetry in which only one plane of symmetry exists. Most leaves exhibit dorsiventral symmetry, as do many flowers. Some leaves, such as those of American elm, show no plane of symmetry. It is interesting to note that in plants where individual flowers or leaves show dorsiventral symmetry or are asymmetric (such as elm), the inflorescence as a whole or the stem as a whole shows a more recognizable pattern of symmetry. An example of this can be seen in figure 4 where two of the individual flowers of a Zinnia head have been removed and each shows dorsiventral symmetry. Figure 5 shows the Zinnia inflorescence which consists of many such flowers arranged in such a way that radial symmetry is observed.

The internal structures of vascular plants also show interesting symmetrical patterns which again are controlled by events initiated in the shoot or root apex. When seen in cross section, the tissues which make up the stem or root show radial symmetry. Figures 6, 7, and 8 show examples of such cross sections.

While symmetries in the plant kingdom are interesting and often aesthetically pleasing, the events which regulate the developmental processes are still to be explained by the student of plant growth and development.

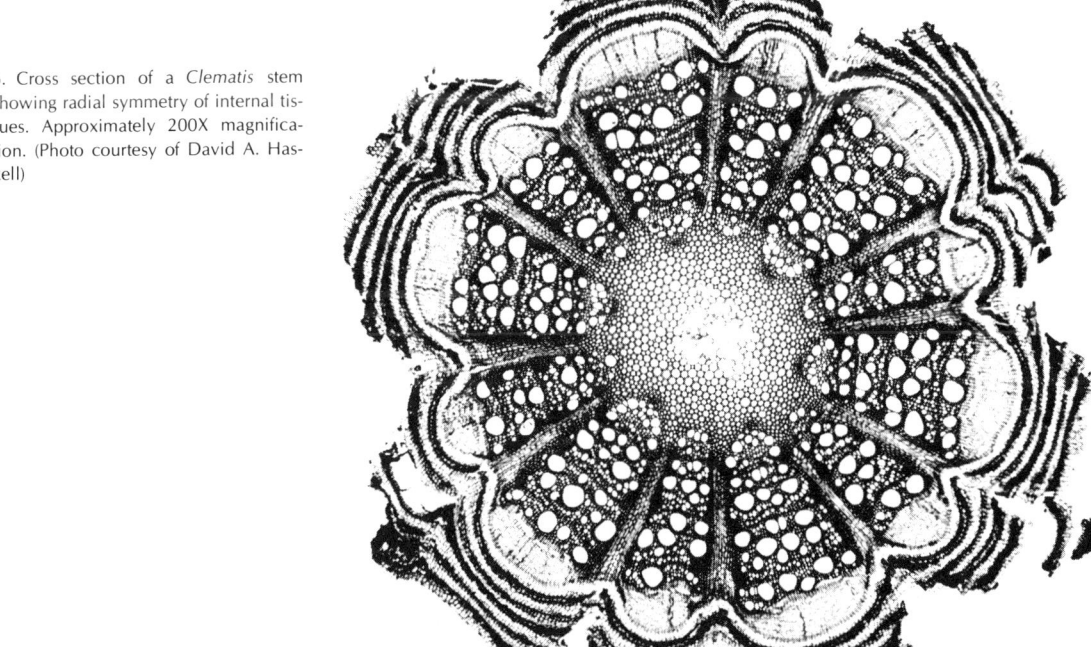

6. Cross section of a *Clematis* stem showing radial symmetry of internal tissues. Approximately 200X magnification. (Photo courtesy of David A. Haskell)

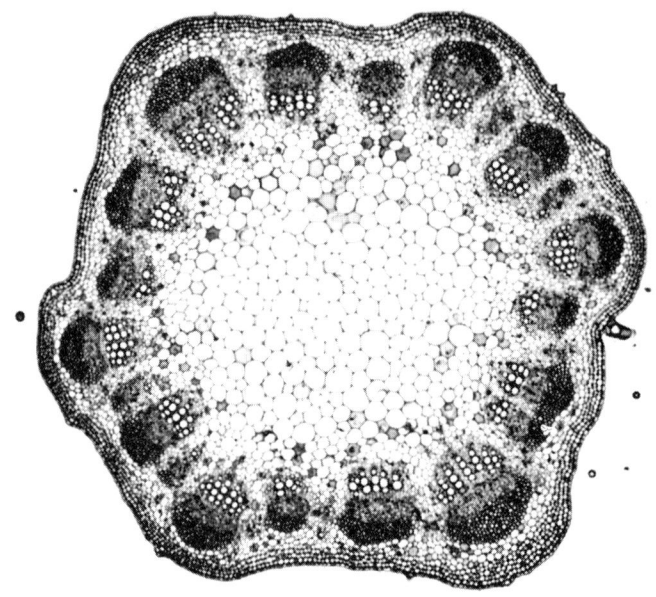

7. Cross section of *Helianthus* stem showing radial symmetry of internal tissues. Approximately 100X magnification. (Photo courtesy of David A. Haskell)

Looking for Symmetries in Plants

Listed here are some home and greenhouse plants that illustrate particular symmetries:

> *Aloe arborescens*—screw axis
> *Davallia canariensis* (rabbit's foot fern)—glide plane
> *Dionaea muscipula* (Venus flytrap)—mirror plane
> *Ficus elastica* (rubber plant)—screw axis
> *Ficus lyrata* (fiddle-leaf fern)—screw axis
> *Phoenix roebelenii* (pygmy date palm)—mirror plane
> *Stapelia gigantea* (Aztec lily)—5-fold rotation

The following plants were selected from the collection of the Smith College Lyman Plant House for display at the 1973 Symmetry Festival:

> *Aloe ciliaris*—screw axis
> *Alpinia formosana*—glide plane

103 Symmetries in Plants

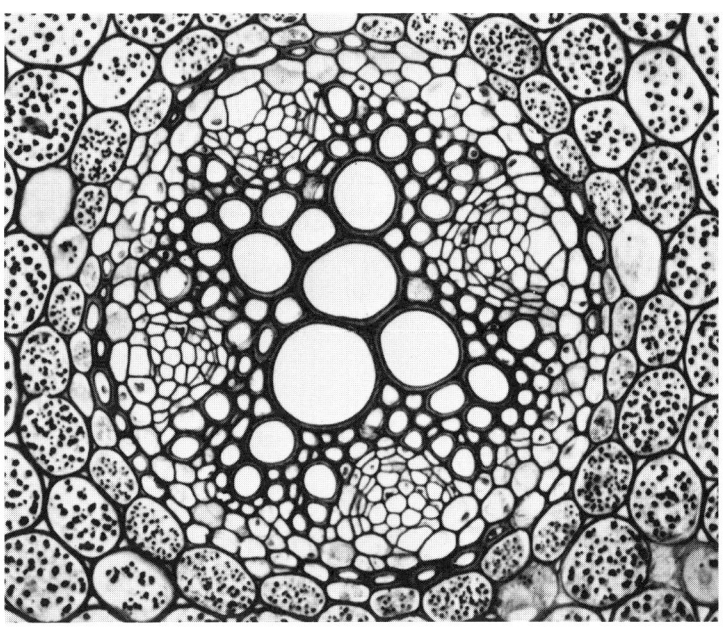

8. Cross section of *Helianthus* root showing radial symmetry of internal tissues. Approximately 200X magnification. (Photo courtesy of David A. Haskell)

Apecra spiralis—screw axis
Astrophyta myriostigma (Mexican star cactus)—5-fold rotation axis
Carludovica palmata (Panama-hat plant)—mirror plane and glide plane
Cymbidium flirtation (an orchid)—bilateral symmetry
Echinopsis oxygona (Brazilian hedgehog plant)—14-fold rotation axis
Hypocryta wetsteini—mirror planes and screw axis
Kalanchoe daigremontiana—mirror plane and screw axis
Kalanchoe tetra-vulcanis—4-fold axis
Maranta leuconeura—leaves show mirror planes
Opuntia speciosa (prickly pear)—mirror plane
Oxalis variabilis (wood sorrel)—leaves form 3-fold axis; each leaf has a mirror plane
Rhoeo discolor (boat lily)—screw axis
Stapelia bigantia (carrion flower)—4-fold rotation axis
Trandescantia reginia—glide plane

GEORGE FAYEN

Ambiguities in Symmetry-Seeking: Borges and Others

By now, in the poet's words, we know "Something there is that doesn't love a wall," and we know too there is something that does. Nature ever reminds us of its necessary opposites. Constantly we are compelled to accept the back-and-forth of things as we watch the contrary movements within our own thinking and feeling. Our minds tend to complicate and yet to simplify, often at the same moment. We want surprise; we want sameness. We seek the open; we seek the closed. We want to make connections; we want to make distinctions. And naturally so: no concept or artifact or organism lives except by an identifying, isolating boundary that still allows it to relate to some ground which is its community and its context. Loving a wall. Not loving a wall.

I

I run the risk of being elementary and abstract because I want to approach symmetry, especially symmetry in literature (insofar as it can be discussed), among other more general human activities. It appears entirely possible in mathematics, the natural sciences, and certain arts to give symmetry a strong, specific definition. In space it results from the reflection of a figure in a mirror plane, or from rotation around an axis or point, or from translation along a line; in time it results from repetition according to a regular pattern. Symmetry can be seen to work both analytically (as a model to examine certain structures in nature) and synthetically (as a module to build works of art). These kinds of precision are not common in literature. Perfect reversals we can find only in the palindrome (Able was I ere I saw Elba), though there are certain kinds of poems with geometric typographical designs, for example, George Herbert's "Easter-Wings" and "The Altar." Even in the working of poetic meter, where the kinship with music might make symmetry more likely, the varying verbal content renders impossible the exact and often simultaneous congruity required by strict symmetry. Various stanzaic and metrical forms can have a roughly symmetrical impact—especially in refrains and incantatory chants where repetitiousness seems to give power. The sestina may perhaps be the closest thing we have to change ringing. But generally the effect of meter and stanza is subtly to avoid symmetry by a constant, playful displacement of expected regularity. "It is this contrast between fixity and flux," said T. S. Eliot, "this unperceived evasion of monotony, which is the very life of verse."[1] Loving a wall; not loving a wall.

It may not be possible in literature, as it is in mathematics or the natural sciences, to find specifically defined elements of symmetry. Rotation and reflection or translation do not appear simply in literary works, although attempts have been made to discover detailed symmetrical designs in some authors (for example, Homer and Dante) who demonstrate sovereign skill and command of large poetic effects. The opening and closing books of the *Iliad* do look to one another. Throughout the epic individual episodes are connected by anticipation and recollection and repeated oral formulae—all in an heroic world where action and reaction dominate the workings of nature and history. But to interpret the *Iliad* in terms of Mycenaen High Geometric art, to divide and set out its book in elaborate ring compositions, is to see symmetry where there is only relationship.[2] It is likewise with Dante's *The Divine Comedy,* which reveals intricate gradation among and within its three parts. It has an elaborately formal organization, but the correspondences, the reappearing types and motifs, and the numerical arrangement of vices and virtues are not merely repetitive. Rather, they are figure and fulfillment and tend towards a final, inclusive image. There is movement in Homer and in Dante, and it is exquisitely formal, but there is nothing we can call rotation or reflection or translation in a symmetrical sense.

The difference of language from matter or visual materials or physical motion indicates other difficulties in finding strict symmetries in literary works. In mathematics, the natural sciences, and in certain arts—whether the figure be equation or crystal or cell or musical score—the symmetry is essentially complete. It may evoke but it does not point, as words point, so richly elsewhere; it does not signify connotatively. Hence in Homer, Virgil, and Dante, as in Shakespeare and Milton and any great writer, it will be difficult to conceptualize any correspondences apart from the forward moving, vital force and flow of the narrative. Considerable efforts, however, have been made of late to see literary works in space and to discuss their "spatial form." These may assist, as a simile might assist, to perceive a work in its whole plot outline, but we are ultimately left outside the work. We experience it less as a process and more as an object; it is observed as an independent structure, a form in space.

What we lose with this approach, and what we can lose with the inevitable limitations of strict symmetry, is the sense of how literature exists in time—in the very personal time of our reading or hearing or in our watching a drama before us or in our mind's eye. This distinction between spatial and temporal art is antique, at least as old as Lessing's

Laocoön (1766), but valid still in its applications. It enables us to ask whether any element in literature can ever be truly identical (or symmetrical) to an earlier one because it must, coming later in time, inevitably be perceived through its remembered likeness and stand itself in a different context. While this blurring will prevent any exactness of symmetry, nevertheless it can act to create a real poignancy and power.

The same face, the same gesture, the same locale or circumstances at a later time: these can evoke memories and enter into the self's recollection of its own nature. From Dante in *The Vita Nuova* through Wordsworth in *The Prelude* and Proust in *A La Recherche du Temps Perdu* and Joyce in *The Portrait of the Artist,* the congruity/incongruity of a *now* and *then* has supplied a measure for growth. Myths of return and repetition abound in the psychology of our experience as well as in our historiography. On this subject Kierkegaard's *Repetition* and Eliade's *Cosmos and History* offer extended meditations. Often we do grow by such rhythms of remembering and recurrence. At best these can look ahead of fulfillment.

> Full souls are double mirrors, making still
> An endless vista of fair things before,
> Repeating things behind.
> (George Eliot, *Middlemarch,* epigraph, ch. 72)

But sometimes, too, we need to express, in the movement of our bodies and minds or the shape of our artifacts, the tendency we find in nature to turn back upon itself or reflect upon itself.

Human experience (identity as our consciousness of continuity in time) so often demands the reflective and the reflexive: so much of what happens to us tends to be curiously binary—as if our epistemology had its own metrics. At such times the prefix *re* comes to dominate the verbs; the activity in time is *again,* and in space is *alike*. Episodes may so come to resemble one another that they approach momentary congruence and similitude: parody and analogue, phases and cycles. People similarly may move through resemblance toward apparent identity: other selves and moments of mirroring, déjà vu and déjà entendu. All these sudden recollections involve a kind of doubling which roughly implies certain workings of symmetry. In this sense, while strict congruence is impossible spatially, the symmetrical can remind us of some of the varied kinds of coherence available in literature.

In mathematics and the natural sciences, as earlier essays here demonstrate, symmetry can do much more. Its immense value is not primarily in ideal theoretical models but in what these models allow us to discover in the coarser, more proximate world of matter. In chemistry, the principles of symmetry can be a guide to imagining the outlines of molecular structure we cannot see. In biology, researches into viruses are proceeding within limits determined by the symmetrical qualities of certain shapes. In geology, movements in the earth's crust are detected by the symmetrical arrangement of anomalies in magnetic lines and along midocean ridges. It is uncertain, though, whether we can approach a work of literature as we can a crystal or plant cell or molecular structure (or even a musical score or dance pattern). The danger here might be in becoming briefly like Sir Thomas Browne, the seventeenth-century naturalist, who wrote in endless detail (*The Garden of Cyrus*, 1658) about finding the quincunx (:·:) everywhere. Is there, after all, in a literary work some essential central structure which, once determined in the form of its basic units, will reveal the larger structure of the whole? Can there be a working model separate from the substance in which the work is realized? Whatever the larger patterns of literary coherence, these are not likely to mark out the course of its growth. A work of words comes into being and develops quite differently from an organic or inorganic structure or even a work in sound or motion. Literature is neither modular nor often open in its essentials to model making.

II

What we can locate in literature are certain impulses toward symmetry and certain symmetrical effects which enter obliquely into the action. In such cases the symmetry involves activities of mind and motives within the narrative by the author. Let us look, as an example, at a short story by the Argentinian writer Jorge Luis Borges from his collection *Ficciones*. Called "Death and the Compass," it comments remarkably on the ambiguities in pursuing symmetries. The first three sentences of the story set the tone:

> Of the many problems which exercised the daring perspicacity of Lönnrot none was so strange—so harshly strange, we might say—as the periodic series of bloody acts which culminated at the villa of Triste-le-Roy, amid the interminable odor of the eucalypti. It is true that Erik Lönnrot did not succeed in preventing the last crime, but it is indisputable that he foresaw it. Nor did he, of course, guess the identity of Yarmolinsky's unfortunate assassin, but he did divine the secret morphology of the vicious series as well as the participation of Red Scharlach, whose alias is Scharlach the Dandy.[3]

The implications here, hauntingly specific (as if axiomatic and pervasive everywhere), gather around Lönnrot's partial success and the ominous incompleteness ("foresaw" not "prevent") of his astutness. Lönnrot emerges sharply as a "pure thinker," the "reasoner" who, when faced with the facts, observes to Police Commissioner Treviranus "that reality may avoid the obligation to be interesting, but . . . hypotheses may not" (p. 130).

The middle of the story finds Lönnrot traveling south toward the villa of Triste-le-Roy. Already the facts involve several murders—the first of a rabbi, Doctor Marcel Yarmolinsky, delegate from Podolsk to the Third Talmudic Conference, stabbed in his room at the Hotel Nord on December 3. One clue: a typewritten scrap—"*The first letter of the Name has been spoken.*" Among the Rabbi's books Lönnrot had begun studying was a monograph on the Tetragrammaton. One month later, on January 3, in the western suburbs, the body of a petty criminal was discovered—and on the yellow and red painted rhombs of a nearby wall, the words "*The second letter of the Name has been spoken.*" Exactly one month later, on February 3, an informer, Gryphius, was taken from his tavern lodgings by two short masked harlequins (yellow, red, green rhombs on their costumes) and driven away—leaving nothing but the scrawled words: "*The last of the letters of the Name has been spoken*" (pp. 131, 133, 134).

Lönnrot finally sees the solution in an anonymous letter and map forwarded to him, promising no fourth crime because the sites of the other three were the vertices of a perfect equilateral triangle. "Symmetry in time (the third of December, the third of January, the third of February), symmetry in space as well . . . to decipher the mystery . . . the word 'Tetragrammaton?'" (p. 136). With the fourth point charted to form a rhomb on the map Lönnrot travels south to the villa. Inside, his steps carry him up a spiral staircase to an observatory. "The evening moon shone through the rhomboid diamonds of the windows, which were yellow, red and green. He was brought to a halt by a stunning and dizzying recollection" (p. 138). Too late: two short men seize him, and Erik Lönnrot is face to face with Red Scharlach, the criminal who blames Lönnrot for his own wound and the arrest of his brother. Scharlach explains the periodic series of crimes.

> I interspersed repeated signs that would allow you, Erik Lönnrot, the reasoner, to understand that it is *quadruple*. A portent in the North, others in the East and West, demand a fourth portend in the South; the Tetragrammaton—The name of God, JHVH—is made up of *four* letters; the harlequins and the paint shop sign suggested four points . . . I sent the equilateral triangle. I sensed that you would supply the

> missing point ... which would form a perfect rhomb, the point which fixes where death exactly awaits you. In order to attract you I have premeditated everything, Erik Lönnrot, so as to draw you to the solitude of Triste-le-Roy. (Pp. 140–141)

So much for the "reasoner." Death follows almost at once.

Borges' story widens out unexpectedly at the end in a way which relates its symmetry-seeking to his other work. Lönnrot, speculating on "symmetric and periodic death," imagines a future encounter. "In some reincarnation when you hunt me," he tells Scharlach, along a single-line labyrinth A to B with C and D as successive midpoints, "kill me at D, as you are now going to kill me at Triste-le-Roy" (p. 141). This prospect of recurrence or widening possibilities opens up the ending of countless Borges' fictions. The dreamer in "The Circular Ruins" (like Erik Lönnrot, far less in control than he thinks) finds that he may himself be someone else's dream. So often in Borges, he who believes himself an agent discovers himself instead to be the instrument of some other agent, who discovers himself in turn ... and so on. The danger seems to be in trying to orient oneself by any one particular point—not simply in seeking symmetry, but in any attempt by the mind at an inclusiveness (be it library, lottery, encyclopaedia, or life work) which fails to acknowledge what is excluded. In "The Moon" the man seeking to make an abridgement of the universe in a single volume neglects a "burnished disc" in the night sky: like the moon, the "essence is always lost." Something beyond may be imminent.

Sometimes, though, the beyond appears to be more of the same. The story may seem about to repeat itself in ramifications, like a module, like an element in the translation or the rotation of a symmetric design. Decisions in "The Babylonian Lottery" are diverging endlessly into others; nothing is final. The worlds of *April-March* regress and ramify into a rage for symmetry in "An Examination of the Work of Herbert Quain." Hexagonal galleries multiply and add on and on to comprise the universe in "The Library of Babel." In "The Garden of Forking Paths" is conceived an infinite network of times, "the strands of which approach one another, bifurcate, intersect or ignore each other through the centuries"; they "embrace *every* possibility." For the agent and his victim "time is forever dividing itself toward innumerable futures" (p. 100).

This ramifying and spiraling outwards from the story into future variations may both frighten and fascinate Borges, who sees it is as the inevitable curse of any image maker. "As a child, I felt before large mirrors that same horror of a spectral duplication of multiplication of reality."[4] Hence, perhaps, Borges' uncertainty about the art of writing: endless possibilities may be mere reflections and offer only claustrophobic mimicry and the pursuit of replicas. Perhaps this is involved in Borges' desire to work back through genealogies to heroic figures, or to objects that are themselves and not their meanings ("Cross, lasso, and arrow—former tools of men, debased or exalted now to the status of symbols"), or to find words which are transcendentally alive.

> Tomorrow they will live again,
> Tomorrow *fyr* will not be *fire* but that form
> Of a tamed and changing god
> It has been given to none to see without an ancient dread.[5]

None of these (hero, tools, *fyr*) can be divided or translated readily into a likeness or secondary symmetric replica. Each is original.

III

More important implications for symmetry-seeking are suggested in "Death and the Compass" by the two central figures and the "desolate symmetrical" villa of Triste-le-Roy. In outlook and manner, as in name, Erik Lönnrot and Red Scharlach seem adversaries, each of whom reflects the other. At the end Lönnrot felt an "impersonal, almost anonymous sadness" (p. 138). Scharlach was "indifferent . . . a fatigued triumph" in his voice (p. 141). What for Lönnrot was a problem was for Scharlach a deadly game; the "periodic series" of crimes was for each the obverse of the other's design. The encounter at the villa seems a confrontation of mind by its antagonist, more the meeting of counterparts and doubles than men engaged in simple conflict. Like any detective story, it offers too the click of closure as question finds congruent answer and mysteries converge on resolution.

It is Scharlach's vengeful manipulating which bears most directly here: it suggests that the revenge hero may as a type be peculiarly involved in symmetrical thinking. Resolute singlemindedness in Red Scharlach took the form of exploiting Lönnrot's overconfidence in his analytic, pattern-finding powers. He chose as trap a caricature of his quarry's weakness: the villa of Triste-le-Roy "was seen to abound in superfluous symmetries and maniacal repetitions" (p. 137). Scharlach confronted his enemy with an image of the simplistic patterning into which Lönnrot's intellect had momentarily stiffened.

We find this reflexiveness in the central scenes of many revenge dramas, where it often becomes an act confronting the victim with his own crime or forcing him (symmetrically, almost as an inversion) to reenact it or to gaze upon its image. Two instances—and I offer these sketchily for my readers to qualify and refine—must suffice. At a crucial moment in the first and second part of Aeschylus' *Oresteia* an avenger confronts a victim; each time the victim is brought into relationship with the deed for which punishment is about to be given. In *Agamemnon* the king must for Clytemnestra walk upon the crimson carpet, thus trampling down the "delicacy of things inviolable" (ll. 370-71) and recalling the "saffron mantle" (l. 239), Iphegenia's blood.[6] Bound with that carpet, soon through that carpet he will be stabbed. Clytemnestra, the avenger in *Agamemnon*, becomes the victim in the *Libation Bearers* when she is confronted with a stranger (her son Orestes—in effect Agamemnon's ghost) seeking hospitality at her door just as once earlier she was confronted with her husband at the door expecting welcome. Clytemnestra avenging the sacrifice of Iphegenia on Agamemnon, Orestes avenging the murder of Agamemnon on Clytemnestra: each time the avenger manipulates circumstances so that the victim will experience something he inflicted. *Lex talionis*. It is action-reaction, systole and diastole—and it is this endlessly recurrent cycle of avenger-victim (curse as symmetry) that the trial in the *Eumenides* (antagonists now adversaries in law) will try to resolve in its symbolic and social harmonizing of male-female, light-dark, and the other polarities which have divided this world.

The revenge in *Hamlet* shows similar aspects of symmetrical thinking. The punishment must in truth fit the crime. For a brief moment Hamlet thinks of killing Claudius at prayer:

> And so I am revenged. That would be scanned.
> A villain kills my father, and for that
> I, his sole son, do this same villain send
> To Heaven.
> Oh, this is hire and salary, not revenge.
> He took my father grossly, full of bread,
> With all his crimes broad blown, as flush as May . . .
> . . . Am I then revenged,
> To take him in the purging of his soul,
> When he is fit and seasoned, for his passage?
> (III, iii, 75-81, 84-86)

Hamlet would rather catch the conscience of the king by reflecting his crime mimetically in the play and by holding up before Claudius' eyes (as he held up the locket before Gertrude's eyes) a mirror image of what has been done. It is this directorial aloofness, this desire to let things reflect upon themselves and apart from his acts work out their intended end—it is this indirect action, this essential inaction which Hamlet has abandoned by the graveyard scene. No longer is he the manipulator or director of playing.

> Our indiscretion sometime serves us well
> When our deep plots do pall . . .
> Ere I could make a prologue to my brains,
> They had begun the play.
> (V, ii, 8-9, 30-31)

Unpremeditatingly he moves, and he begins at last essentially to act.

I have digressed briefly from Borges to suggest that certain aspects of revenge in "Death and the Compass" can be found in many other literary works and that these illustrate certain crucial qualities in symmetry-seeking. As a type, the avenger is of course an *alazon*, an overreacher striving for more than is allowed, and yet is as well a figure from comedy.

> As when some one peculiar quality
> Doth so possess a man, that it doth draw
> All his effects, his spirits, and his powers
> In their conflutions, all to run one way,
> This may be truly said to be a humour.
> (Ben Jonson, *Every Man Out of His Humour*, 1599)

Singlemindedness is at the center of certain comic types who continue happily to repeat themselves forever. Lawrence Sterne in *Tristram Shandy* (1760–67) has given us two eccentrics afflicted by a "humour." Uncle Toby must find out where he was wounded; this involves making interminably a model of the outworks of Namur. Walter Shandy wants to direct his son's education by an elaborate program, the *Tristrapaedia*, whose composition proceeds less rapidly than the growth of the boy it is meant to instruct. Both men, Bergson would remind us (*Le Rire*, 1900), are mechanical. But both are also symmetry-seeking in that they want time to conform quite exactly to the outline of their schemes. In wanting to control reality they repeat themselves endlessly. When not simply comic, such singlemindedness displays an intensity and a peculiar hyper-attentiveness in reaching for control. Looking backwards, it demands that the present be symmetrical to the past. Looking forwards, it will demand that the future be symmetrical to the present. More precisely, it will assume that certain moves in time and space will result in configurations congruent to those already taken by matter or mind.

We can regard the villa of Triste-le-Roy in Borges' "Death and the Compass" as a monument to what can happen when these demands are made in earnest, when problem-solving becomes symmetry-seeking and then turns into model-making and system-building. At the villa a "glacial Diana in one lugubrius niche was complemented by another Diana in another niche; one balcony was repeated by another balcony; double steps of stairs opened into a double balustrade" (p. 137). The ultimate condition implied by this state of mind can be seen initially in some instances from other literary periods. Closed repetitiveness and doubling abounds in Timon's villa, cited for its false magnificence in one of Pope's Moral Essays.

> No pleasing Intricacies intervene,
> No artful wildness to perplex the scene;
> Grove nods at grove, each Alley has a brother,
> And half the platform just reflects the other.
> The suff'ring eye inverted Nature sees,
> Trees cut to Statues, Statues thick as trees.
> (*Epistle IV*, To Burlington, ll. 115–20, 1731)

Versailles-like in its extravagant balance, it does not know the secret of an English garden.

> He gains all points, who pleasingly confounds,
> Surprizes, varies, and conceals the Bounds.
> (ll. 55–56)

These features in the landscape ("pleasing intricacies," "artful wildness," the varied "Bounds"), along with the "good sense" which Burlington cultivates in things natural and human, will release and truly refresh. Together they constitute Pope's alternative to Timon's symmetries.

This geometric and mechanical style can be found in the Newtonian universe so detested by William Blake, detested for regularity and its repressive rationalizing of motion. The god of this universe was Nobodaddy, and its society offered "mind-forg'd manacles" ("London," *Songs of Experience*). Against these prohibitions Blake will set lyrics of paradox and inversion. In "The Tyger" he asks ironically

> What immortal hand or eye
> Could frame thy fearful symmetry?

But to the speaker (who sees not through but with the eye) the tiger appears awesomely formed. This view would be according to the "ratio of all things" in which we can only "repeat the same dull round over again." The "universe would soon become a mill with complicated wheels" ("There is No Natural Religion," I & II).

1. M. C. Escher, *Swans*, wood-engraving (1956). (Reproduced by permission of the Escher Foundation, Haags Gemeentemuseum, The Hague.)

Such a repetitive symmetry Blake rejects for release and movement. "Without contraries is no progression," he proclaimed in *The Marriage of Heaven and Hell.* "Energy is the only life and is from the Body and Reason is the bound or outward circumference of Energy." These contraries Blake saw embodied in certain figures.

> The Giants who formed this world into its sensual existence, and now seem to live in it in chains, are in truth, the causes of its life, & the sources of all activity, but the chains are the cunning of weak and tame minds, which have power to resist energy . . .
>
> Thus one portion of being is the Prolific, the other, the Devouring: to the devourer it seems as if the producer was in his chains. . . .
>
> But the Prolific would cease to be Prolific unless the Devourer as a sea received the excess of his delights.

This kind of relationship between Devourer and Prolific must come to replace the "dull round" and the "mill." Here and in Blake's other works, like "The Mental Traveller" and the Prophetic Books, the rhythms of renewal and sexual reciprocity are his reaction against symmetry-seeking raised to a higher philosophic power.

The symmetries rejected by Pope and Blake (the one for "artful wildness," the other for lack of vital energies)—whether in architecture, gardening, or metaphysics—were each flawed by obsessive design. Visually it would be in the effect of a closed system where everything balances and each part corresponds to another part, all in mirror images and reversals: like M. C. Escher's engraving "Swans" (fig. 1), those black-and-white shapes seeming to fly both ways simultaneously, each the ground for the other's figure, so that it all fits together completely, symmetrically.[7] We get something of the same sort of literary works which seem overdesigned, where episodes tend to repeat and certain settings or characters keep reappearing in geometric patterns almost obsessively, sometimes to such a degree that these repetitions seem less the author's design than the outward and visible symptom of characters whose own inner need for certainty and order expresses itself in sharp rectilinear outlines.

These characters keep doing the same things, and this repetitive mechanical behavior may be accompanied by certain kinds of stylizing and abstract language where we find a loss of particularity and of trust in the personal. The life style which accompanies this withdrawal may tend toward mental economy in people who lack enough energy to be tolerant or flexible or hospitable to anything new.[8] Their rigidities and abstraction come from fatigue or fear; they become narrowed in habitual protective patterns. Thomas Hardy and D. H. Lawrence have characters of this kind, especially in *Jude the Obscure* and *Women in Love,* and their own fictional techniques reveal in language and structure this stress on obsessive behavior.

Kafka particularly gives us in his stories people who are locked within their own repetitive systems of anxiety. In "The Burrow," no matter who or what the speaker is, one feels he will continue to scuttle back and forth from the Castle Keep to the surface, ever seeking compulsively to make his domain perfect and complete. Such a predicament, immobilizing as in Beckett, will operate like the "double bind" formulated by R. D. Laing (*Knots*). Each of the alternatives will neatly cancel the other, as they do also in Kafka's kind of paradox. "The Crows maintain that a single crow could destroy the heavens. Doubtless that is so, but it proves nothing against the heavens, for the heavens signify simply: the impossibility of crows."[9] The two halves of this statement coexist and yet crisscross simultaneously like the swans which in Escher's engraving fly both ways at once: optical illusion as a spatial rendering of paradox. Witty, to be sure, but despondent and with no way out. The statement, the engraving, and the story are utterly closed.

Another way of describing what may be happening when symmetries become closed and stylized is to say that the differences needed to make the similarities significant are no longer strong and clear enough to allow for vital discrimination. It is as if the mind had moved from assimilating what is outward to completing relationships so inward, so inturned and exclusive that the work becomes autonomous in ways which abandon all relationships and even the very nature of relationship itself. Recent studies suggest in schizophrenics an impairment in the ability to relate and construe one thing in terms of another.[10]

This brief excursion into the pathology of symmetry should not obscure the fact that symmetry-seeking is only a special case of the human desire for proportion and repetition, special in being more rigorous and geometric, special also in satisfying a far greater need for strict inner equilibrium. Its problems are the familiar ones magnified. We still do not know exactly how proportion is involved in body image or in the sense of habitable space or territory. We still do not know fully how repetition contributes to self-consciousness or, through rites of communal memory, how it informs the consciousness of a culture. Clearly our individual needs for symmetry differ widely, but what we do know is the psychological importance of asymmetry and irregularity and disproportion.[11] Snowflakes in Thomas Mann's *The Magic Mountain* take on an ominous perfection:

> ... each of them was absolutely symmetrical, icily regular in form. They were too regular ... the living principle shuddered at this

> perfect precision, found it deathly, the very marrow of death—Hans Castorp felt he understood now why the builders of antiquity purposely and secretly introduced minute variations from absolute symmetry in their columnar structures.[12]

One point certain is the value of coming more to understand how symmetry and asymmetry relate in affecting the way we live. The right hemisphere of the brain mysteriously does not correspond to the left.

Ambiguities and dangers remain. Symmetry in its etymology means "with measure." To look for measured order and for connections among disparate areas is well and good; it is our collective destiny. Symmetry may even help us resolve or at least rephrase the dilemma of the "world-knot."[13] But even now the warning of Sir Francis Bacon should be heeded.

> The human understanding is of its own nature prone to suppose the existence of more order and regularity in the world than it finds. And though there be many things in nature which are singular and unmatched, yet it devises for them parables and conjugates and relatives which do not exist.
> (*Magna Instauratio,* Aphorisms xlv)

Models may be essential to the construction of predictive explanations, but like any system of representation they may come to replace the reality they were meant to serve: vision may harden into dogma. We must respect what is singular.

Through symmetry we know we can discover a norm for many things, a norm to which things will in their own way approximate their deeper and generic nature. When does this search begin to tend toward excess, toward a need for proportion and repetition and regularity so acute that it becomes obsessive and life-denying? How do we know when symmetry-seeking is still human? Perhaps only when it doesn't love a wall.

LOREN EISELEY **Notes of an Alchemist**

Crystals grow
      under fantastic pressures in the deep
                      crevices and confines of
the earth.
      They grow by fires,
           by water trickling slowly
in strange solutions
      from the walls of caverns.
They form
           in cubes, rectangles,
                tetrahedrons,
                      they may have
their own peculiar axes and
           molecular arrangements
                but they,
        like life,
           like men,
               are twisted by
the places into which
           they come.

I have only
      to lift my hands
           to see
      the acid scars of old encounters.
                    In my brain
as in the brains of all mankind
              distortions riot
and the serene
      quartz crystal of tomorrow is
most often marred
      by black ingredients
           caught blindly up,
     no one knows surely why
            specific crystals meet
in a specific order.

"Notes of an Alchemist"

                    Therefore we grasp
two things:
                    that rarely
two slightly different substances will grow
                                        even together
but the one added ingredient
                    will transfigure
a colorless transparency
                    to midnight blue
or build the rubies' fire.
                    Further, we know
that if one grows a crystal
                    it should lie
under the spell of its own fluid
                              be
                    kept in a cool cavern
                         remote
                              from any violence or
                                        intrusion from the dust.

So we
        our wise men
             in their wildernesses
                              have sought
to charm to similar translucence
             the cloudy crystal of the mind.
We must then understand
             that order strives
against the unmitigated chaos lurking
along the convulsive backbone of the world.

Sometimes I think that we
             in varying degrees are grown
like the wild crystal,
             now inert,
now flashing red,
        but I
             within my surging molecules
by nature cling
        to that deep sapphire blue
                    that marks the mind of one

long isolate
        who knows and does reflect
starred space and midnight,
            who conceives therefore
that out of order and disorder
                  perpetually clashing and reclashing
come the worlds.
        Thus stands my study from the vials and furnaces
of universal earth. I leave it here
                for Heracleitus
if he comes again
        in the returnings of Giant Year.

4 A FINAL FOCUS

1. The famous international *Belvedere Center for the Study of Symmetry in the Netherlands*. (M. C. Escher, *Belvedere*, lithograph, 1958. Reproduced by permission of the Escher Foundation, Haags Gemeentemuseum, The Hague.)

MARJORIE SENECHAL

The Theory of Patterns

"My brain already reels. My mind, like a cloud momentarily illuminated by a lightning flash, is for an instant filled with an unusual light, which now beckons to me. . . ."
Galileo

I. Prologue

Time: the present. A warm and sunny afternoon
Place: the courtyard of the Belvedere Center for the Study of Symmetry

Festivant Good afternoon. I am looking for Professor Symmetrica Smith, who teaches at this Center. Do you know where I can find her?

Symmetrica I am she, but please call me by my first name. We are quite informal here. You must be Festivant. I just received your letter this morning. Do sit down. You must be tired, after climbing up this steep mountain.

Festivant I am worn out, indeed. Why isn't there a road leading to the summit? I had to climb over huge rocks and through brambles. I lost my way several times.

Symmetrica No one has been able to build an easy way up this mountain. The best engineers have studied it from every perspective; they say it's hopeless.

Festivant Then why did you build the Center here? And why such a magnificent tower, when so few will ever climb it?

Symmetrica Because the view from the top is an incomparable one.

Festivant May I climb to the top? I would like to see it.

Symmetrica No, not yet. You must wait until you are better prepared to appreciate it. Now, what can I do for you? In your letter, you said you wished to continue your study of symmetry. But you did not explain why, nor what you wanted to learn.

Festivant The reason is simple: I am intrigued. I have read all the articles in *Patterns of Symmetry*. I was astonished to find symmetrical patterns in so many fields—dance, ornament, plants, crystals, literature, music. But when I finished, I realised that there is still much more to learn. For example, there is surely symmetry in architecture, in physics, and in many other fields.

Symmetrica But you will discover these patterns for yourself. Now that your eyes are opened, you will find patterns that no one has shown to you—perhaps you will find some that no one has seen before!

Festivant Then what do you do up here, if you don't search for patterns?

Symmetrica We make it possible for others to do so. Here we study the theory of patterns, trying to find the features that are common to them all.

Festivant   That sounds very interesting. Will you teach me your theories?

Symmetrica   You wrote that you can spend just two days here. I can teach you only the rudiments in such a short time. But it will give you a start, and after you leave, you will be able to continue your studies on your own.

Festivant   Very well. When can we begin?

Symmetrica   This evening, if you like. First you must rest, and have a light supper. It is useless to try to do mathematics when one is tired, or hungry—or too full!

Festivant   Mathematics?

Symmetrica   Why, yes. Mathematics is the natural language of symmetry. It would be more difficult to study it in translation.

Festivant   But I am not a mathematician. Perhaps I shouldn't have come.

Symmetrica   Nonsense! Don't be concerned on that account. The mathematics that you must learn, though perhaps new to you, is not difficult. And we will go slowly.

Festivant   This evening, then. What shall I bring with me?

Symmetrica   Nothing. Just a rested mind.

Festivant   Is that really all? No books? No other equipment?

Symmetrica   No, nothing more for now. Books will come later. Now come with me; I will show you to your room.

II. Systems of Symmetries
Time: that evening
Place: in the study of the Belvedere Center

Symmetrica   I hope that you are rested now and ready to begin work. Do you have any questions to start us off?

Festivant   Yes, I do. I have noticed, in the patterns in this book, that certain combinations of symmetry elements frequently recur. For example, where there are two intersecting mirrors, there is always a rotation axis; where there are two intersecting rotation axes there is always a third. I wonder if this is particularly significant.

Symmetrica   It is of fundamental significance! This is why the experienced symmetry-seeker is able to grasp an entire complex of symmetries very rapidly. He has learned to recognize symmetry elements and he knows the symmetry systems they generate. You too should learn to operate with the combinations skillfully and quickly.

Festivant I'm not sure how to go about that.

Symmetrica We can begin by making up a shorthand for them. For example, we can denote symmetry operations (and the corresponding symmetry elements too) by letters, such as *r* for rotation, *t* for translation, and *m* for reflection. Then we can regard their combinations as a sort of arithmetic.

Festivant That idea was mentioned in the introduction, but I don't think I really grasped it.

Symmetrica Then let us review a moment. Suppose we consider reflections in two mirrors. We can use subscripts to distinguish their symbols: m_1 and m_2. Then by writing $r = m_2 \cdot m_1$, where the symbol "·" means "followed by," we can express the fact that whenever two mirrors intersect, rotational symmetry is produced. The operation $m_2 \cdot m_1$ is called the product of m_1 and m_2. The equation, translated into English, says "reflection in m_1, followed by reflection in m_2, equals rotation." Notice that we read the symbols in a product from right to left.

Festivant I like the general idea, but I'm not sure I really understand it. I find it strange, calling *r* and $m_2 \cdot m_1$ equal. In one case, you are rotating, and in the other, you are reflecting in two mirrors.

Symmetrica The use of the word "equals" here is a mathematical convention. It is customary to say that two symmetry operations are equal if they produce the same end result on a motif. I'll show you what I mean. First, I draw two lines to represent the mirrors. Then I place a motif near one of them and label it *a*. Reflecting the motif in m_1 we obtain a second motif *b*. Now reflect *b* in m_2: we get motif *c*. Thus we obtain *c* from *a* by the operation $m_2 \cdot m_1$. But notice: we could also go from "*a*" to "*c*" by rotation about the axis formed by the intersection of the mirrors. In this sense, $m_2 \cdot m_1 = r$.

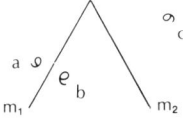

Festivant That's clearer now. But in what ways are symmetry operations analogous to arithmetic?

Symmetrica We can treat them almost if they were numbers; that is, they obey most of the rules of multiplication. First of all, the product (successive application) of any two symmetry operations for a given figure is always a third symmetry operation for it. Second, this product is associative: that is, if m, n, and p are any three symmetry operations, then $(m \cdot n) \cdot p = m \cdot (n \cdot p)$. It's just like multiplication of real numbers, where $(xy)z = x(yz)$. Third, there is a symmetry operation which plays the role of the number 1: it's the identity operation i, which involves no motion at all. Consequently, $i \cdot n = n \cdot i = n$ for any symmetry operation n. And last, if n is any operation, you can both perform it and undo it. The undoing, or reverse operation is called the inverse of n and denoted n^{-1}. We always have $n^{-1} \cdot n = i$, and $n \cdot n^{-1} = i$. This is analogous to the algebraic equation $x(1/x) = 1$; $1/x$ is x^{-1}.

Festivant I'm really quite surprised at this. I would never have thought that symmetry was so closely related to algebra.

Symmetrica Watch out though! There are some differences, too. The product of two symmetry operations is, in general, not commutative.

Festivant Do you mean that the order in which you perform them matters; for example, that $m_2 \cdot m_1 \neq m_1 \cdot m_2$? But why? Aren't both these products rotations?

Symmetrica Yes, and they are rotations through angles of the same number of degrees, but one is clockwise and the other counterclockwise. I'll prove it to you by drawing another figure. The mirrors m_1 and m_2, and the motif a are placed as before. But this time we perform m_2 first, and then m_1.

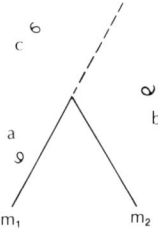

Festivant Now I see.

Symmetrica The mirrors in both diagrams are 60° apart, and the rotations produced by the successive reflections are through angles of 120° and −120° (−120° is the same as +240°). Those two diagrams show only some of the symmetries implied by the mirrors, however. Let's finish the picture. Now we have six symmetry operations, and six motifs; each motif illustrates where a given operation takes the first one, a. Now you can find more algebraic relationships. For example, starting with a, $r \cdot r$ brings us to motif e, and $r \cdot r \cdot r$ brings us back to a again. So $r \cdot r \cdot r$, or r^3 for short, is equal to the identity operation i. Where does $r \cdot m_1$ take motif a? Remember that r is counterclockwise.

The Theory of Patterns

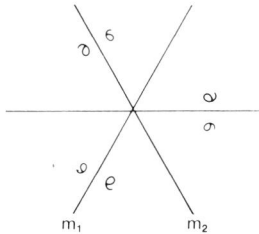

Festivant To motif d. And $m_1 \cdot r$ takes a to f. But don't ask me any more questions, Symmetrica! This is fascinating, but I haven't absorbed it all yet. There is so much to remember. I must try to organize all this information.

Symmetrica There is a very simple way of organizing it. Notice that $m_1 \cdot r$ acts like a third mirror, which we'll call m_3. Thus the symmetry system we have been studying has six symmetry operations, one corresponding to each motif in our diagram. We can display everything we know about this system in a table, similar to a multiplication table in arithmetic. The table will have six rows and six columns, like this:

o	i	m_1	m_2	m_3	r	r^2
i						
m_1						
m_2						
m_3						
r						
r^2						

Now let's begin to fill it in. Each entry will be one of our six symmetry operations. To find the entry in, say, column four and row three, you must find the product of the operation at the top of the fourth column and the operation at the beginning of the third row, performed in that order.

Festivant That would be r^2, since if I reflect a first in m_3, and then reflect that image, f, in m_2, I reach motif e.

Symmetrica Excellent! Can you fill in any more entries?

Festivant The first four entries on the main diagonal (upper left to lower right) must be i, because $m_1 \cdot m_1 = i$, and similarly for m_2 and m_3.

And all the entries in the first row will be identical with the column headings, because to get them we just multiply the column heading by i. Similarly, the entries in the first column will be identical with the row headings.

Symmetrica Good—and what is the last entry in the main diagonal?

Festivant I don't know—let me think. It must be r^4. But r^4 isn't one of the six operations.

Symmetrica Yes it is, but it appears in disguise. Remember that $r^3 = i$. That is, if you rotate three times you are back where you started. So rotating a fourth time is equivalent to rotating just once: $r^4 = r^3 \cdot r = i \cdot r = r$.

Festivant Then I think I can fill in the rest of this table.

Symmetrica Good! That is your assignment for tomorrow. Let me end this evening's work by summarizing what we have discussed. The symmetry operations for any object form a set which has a product in many ways analogous to the multiplication of real numbers. The set is closed with respect to this product (the product of any two operations in the set is also in the set), and it contains an identity operation and the inverses of each of its operations. The product must be associative, but it need not be commutative. Such a set is called a "group" in mathematics. This is a technical term. In ordinary speech, one might refer to a group of people and mean the same thing as a set of people. But a mathematical group is always a set with the properties I just listed. A major part of the study of symmetry is the study of symmetry groups. In mathematical language, the group contains most of the essential information about the symmetry system we wish to study. This evening we have examined some features of a symmetry group generated by two mirrors. But you have also learned something about symmetry groups in general. Let us continue tomorrow morning after breakfast.

Good night!

III. Group Generators and Polyhedral Symmetries

Time: the next morning
Place: the study

Symmetrica Good morning! Did you sleep well? Some people find the air here too rarified.

Festivant I find it clear, pure, and refreshing, and I woke up early to complete the table. Here it is! And after I finished it, I went for a stroll on the roof terrace.

o	i	m_1	m_2	m_3	r	r^2
i	i	m_1	m_2	m_3	r	r^2
m_1	m_1	i	r^2	r	m_3	m_2
m_2	m_2	r	i	r^2	m_1	m_3
m_3	m_3	r^2	r	i	m_2	m_1
r	r	m_2	m_3	m_1	r^2	i
r^2	r^2	m_3	m_1	m_2	i	r

Symmetrica The table's fine. Did you see anything interesting up there?

Festivant This mountain top is truly a beautiful spot. I didn't climb the tower, but I did see some of my fellow students. One was totally absorbed in studying a strange-looking shape, and didn't speak to me. Another was looking through the bars of a dungeon. What is he doing in there?

Symmetrica That's not a dungeon, it's our bell-room; six bells are hung there and we ring changes on them. The man in there is Pierre. He is working on a project of his own, analyzing the patterns of light cast on the floor by the gratings, at different hours of the day. He should be finished in a month or two. The other fellow you saw is Louis. He's an architect who is studying here this year. He's fascinated by our tower and wants to design houses and other buildings based on the same principles.

Festivant Is his cube related to the tower somehow?

Symmetrica Yes, they are constructed in essentially the same way. One way to build Louis's cube is this: beginning with an ordinary cube, choose one face to be the top and the opposite the bottom. There are four columns connecting the top with the bottom—two pairs of adjacent edges. Now cross the members of each pair, and then rotate the top ninety degrees. This gives you Louis's cube. The tower is just the same, but with rectangles for the top and bottom, and eight columns—four pairs—instead of four. In fact, if you look closely you'll see that the four inner columns of the tower are arranged just as in Louis's configuration.

Festivant Good heavens! I never would have noticed that myself. Louis and Pierre are much more advanced than I. I have a question about that simple table. I wonder, if I were given the table and asked to find an object with the symmetry group described by it, how would I do it?

Symmetrica You must learn to look for the key information. For example, from the table you learn that m_1 and m_2 are not parallel, since their product is rotation (if they were parallel, it would be translation); and since the table also tells us that $r^2 \cdot r = i$, we know that r is three-fold. This means that the rotation is through 120° and since the angle of rotation is always twice the angle between the mirrors, we conclude that the mirrors are 60° apart. And then we note that every operation in the group can be written as a series of compositions involving only m_1 and m_2.

Festivant Let me see.... I find $r = m_2 \cdot m_1$ in the table, so $r^2 = (m_2 \cdot m_1)^2$. But I don't see how to write m_3 in terms of m_1 and m_2.

Symmetrica We saw, last evening, that $m_3 = m_1 \cdot r$.

Festivant Now I understand: $m_3 = m_1 \cdot m_2 \cdot m_1$.

Symmetrica That's right. The table shows only products of two operations. But a reflection must be the product of an odd number of reflections.

Festivant Why is this?

Symmetrica Reflections convert a motif to its mirror image, while rotations do not. The product of an even number of reflections in intersecting mirrors would thus be a rotation; notice that this is analogous to the multiplication of negative numbers. Now, since every operation can be written in terms of m_1 and m_2, we can say that m_1 and m_2 generate the entire group. So you see that the table tells you that any object which has two mirrors at an angle of 60° and all the symmetries implied by these mirrors (but no others) can be used to illustrate the group. An equilateral triangle would be the simplest example.

Festivant That really is the idea behind the kaleidoscope, isn't it! One starts with two mirrors and suddenly, there is the entire configuration, with all its symmetries. They are generated by the mirrors.

Symmetrica Yes, and the generators characterize the symmetry groups. In practice, one doesn't try to write down a table for every group, because some groups are very large. The groups of plane patterns are infinite, because they contain translations, which can be repeated over and over without ever bringing a motif back to its original place. But if you know the group's generators, and what happens when they are combined, the rest of the operations can be figured out.

Festivant I recall that Arthur Loeb used generators in his talk (pp. 28–43) on plane patterns and color symmetry. I remember that he began with two parallel rotation axes and used them to generate his patterns.

Symmetrica That's right. He was interested in discovering all possible symmetry groups for plane patterns, so he started with two generators, rotations, which he required to combine to form a third. Thus he supposed that the rotations r_1 and r_2 were k-fold and l-fold (this means that $r_1^k =$

i, $r_2^l = i$) where k and l were numbers to be determined, and that $r_3 = r_2 \cdot r_1$. He showed then that the closure property requires r_3 to be m-fold, where m satisfies the relation $1/k + 1/l + 1/m = 1$.

Festivant Then the actual values of k, l, and m are numbers which satisfy this equation. But how is this group theory related to the simple idea of fitting polygons together to form plane patterns? In Mrs. Senechal's essay it was shown (p. 93) that hexagons, squares, triangles, rectangles, and parallelograms could be used to pave a plane, and that every plane pattern must be based on one of these.

Symmetrica Well, if we consider, for example, the solution (6,3,2) and the corresponding plane pattern created by fitting hexagons together (p. 94), we can make the connection. The six-fold rotation axes are located at the centers of the hexagons. The three-fold axes are located at the vertices of the hexagons, and relate the three which meet there. Where are the two-fold axes?

Festivant At the middle of the edges formed by pairs of hexagons.

Symmetrica Yes, and you can see, geometrically, that the two-fold rotations are the products of six-fold and three-fold rotations. If you place hexagons together, three at a vertex, then they have to share edges, and so a two-fold axis arises. And you can see how the (4,4,2) solution corresponds to paving the plane with squares, and so forth.

Festivant So the two approaches are the same.

Symmetrica No, Loeb's equation only describes the rotational symmetries of the patterns. The tiled planes have all the mirror symmetries of the polygonal tiles as well. To go from a pattern with (6,3,2) rotational symmetries to the symmetries of a hexagonal tile floor, you have to add the mirror reflections, and glide reflections* to the rotation group. We'll discuss that later.

Festivant All right.

Symmetrica You may have noticed that Elliot Offner dealt with generators in his lecture too, although the printers whose work he discussed were not thinking about mathematics explicitly. However, the "principle of varied combinability of the units" which he cited (p. 61) is certainly an intuitive recognition of the various ways in which symmetry groups can be generated. Now, I would like to show you that group theory describes the symmetries of polyhedra as well as plane patterns. First note that the regular polyhedra are analogous to polygonally paved planes.

Festivant Yes, I recall that the faces of a regular polyhedron are all identical regular polygons, such as squares, or triangles.

---

* Glide is $t \cdot m$, with the mirror parallel to the direction of translation.

Symmetrica   Yes, the polygons must be identical and themselves regular; that is, all their edges must have the same length, and all their angles the same magnitude. And the same number of polygons must meet at each vertex of the polyhedron.

Festivant   Then the cube is a regular polyhedron, and so is the octahedron, but the cuboctahedron is not, because six of its faces are squares, while the remaining eight are triangles.

Symmetrica   That is correct, and now you will easily see why there are only five regular polyhedra. For, at least three polygons must meet each vertex, and the sum of the three equal angles presented there must be less than 360°; otherwise the polygons would lie flat, or be crumpled. For example, four squares lie flat, as you know since you have been walking about here and have undoubtedly noticed our tiled floors: each square presents an angle of 90°, and four together make 360°. On the other hand, four hexagons couldn't be fitted together without crumpling. Now, if a polygon has *n* sides, then its central angle is 360/*n*, and each vertex angle is 180(1-2/*n*)°:

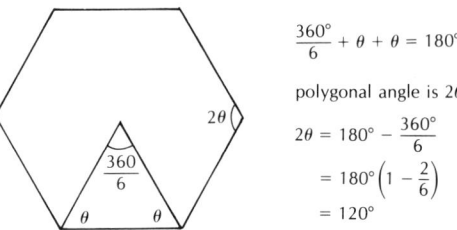

$$\frac{360°}{6} + \theta + \theta = 180°$$

polygonal angle is 2θ

$$2\theta = 180° - \frac{360°}{6}$$
$$= 180°\left(1 - \frac{2}{6}\right)$$
$$= 120°$$

When three of these polygons are put together, the sum of the three vertex angles is 540(1-2/*n*)°; and with a little arithmetic we see that this is less than 360° only if 2/*n* is more than $\frac{1}{3}$, or, what is the same thing, if *n* is less than six. For example, if *n* = 6, the polygon is a hexagon. But three hexagons lie flat, so hexagons can't be used to build a regular polyhedron.

Festivant   Then this means that only pentagons, squares, and triangles can be used.

Symmetrica   Yes, and the number of polygons which meet at a vertex is also limited. Try to fit four pentagons together—the sum of the angles would be 432°, so it can't be done, but three of them, with 324° together, would form a nice polyhedral vertex. The pentagonal dodecahedron has three pentagons meeting at each vertex. Again, only three squares can meet, since four lie flat. Thus the cube is the only regular polyhedron built of squares. But with triangles, there are three possibilities: 5, 4, or 3. The first is the icosahedron, the second the octahedron, the third the tetrahedron.

Festivant So that is why there are only five regular solids.

Symmetrica Yes. Now let us move from the regular solids to their symmetry groups. We can assume that their rotations groups too are generated by two rotations—think of the cube, with its four-fold axes through the square faces, and three-fold axes at the vertices: the two-fold axes arise automatically through the juxtaposition of the squares.

Festivant In a plane pattern, the generating axes are parallel, perpendicular to the plane, while in a polyhedron the axes all intersect in its center.

Symmetrica Yes. Let us begin as Arthur Loeb did, calling the two generators r_1 and r_2, and assuming that they are k-fold and l-fold. And let us say that $r_3 = r_2 \cdot r_1$, and that r_3 is m-fold. We need to determine all possible values of k, l, and m. We can follow Arthur Loeb's argument exactly, except that the diagram (p. 37) must be imagined drawn on the surface of a sphere, instead of on a plane.

Festivant Does this make a difference in the solutions?

Symmetrica Yes, because spherical geometry differs from plane geometry in certain respects. For example, the sum of the angles of a triangle on the surface of a sphere is always greater than 180°, and the sum of the angles of a quadrilateral is greater than 360°. For the quadrilateral in the diagram, the sum of the angles is

$$360/2k + 360/l + 360/2k + 360/m.$$

In order for this sum to be greater than 360, we must have

$$1/k + 1/l + 1/m > 1.$$

Thus we obtain a restriction analogous to that for the plane. But it isn't quite the same; it's an inequality instead of an equality, and so of course the solutions are different. What sets of integers k, l, m satisfy this condition?

Festivant We could have $k=5$, $l=3$, $m=2$, since $\frac{1}{5} + \frac{1}{3} + \frac{1}{2} = \frac{31}{30}$, which is greater than 1. These are the rotations of the pentagonal dodecahedron, and of the icosahedron.

Symmetrica Yes, that's one solution. If r_k is five-fold and r_l is three-fold, where is the axis of the two-fold rotation r_m?

Festivant All the two-fold axes bisect opposite edges, and relate the two pentagons which meet at these edges.

Symmetrica You are thinking of the pentagonal dodecahedron when you say that, but other polyhedra with different sorts of faces can have the same rotational symmetries. It would be better to say that the two-fold rotations relate two adjacent five-fold axes, and two adjacent three-fold axes. Then you are speaking of any polyhedron with these symmetries, rather than a particular one. What are the other solutions?

Festivant   There's also $k=4, l=3, m=2$. This is the group of rotations of the cube—and the octahedron, and many others. Finally there is $k=3, l=3, m=2$. I recognize these as the rotations of the tetrahedron.

Symmetrica   That is very good, but the list of solutions is not yet complete. The solutions you have found correspond to the rotations of the regular solids and indeed there are no other solutions if both k and l are greater than 2. But in addition to these groups, we have the solution $(k,2,2)$ where k can be any integer greater than 1. For example, consider a square sheet of metal. The k-fold axis (here $k=4$) is the axis of the square, and the two-fold axes lie in the plane of the square and turn it over.

Festivant   What about the solution $(k,l,1)$? That satisfies the inequality as well, since $1/k + 1/l + 1/1$ is always greater than 1.

Symmetrica   In this case, r_1 and r_2 are rotations about the same axis. I will explain it to you this way: if $m=1$, then $r_3 = i$, and $r_2 \cdot r_1 = i$. This means that $r_2 = r_1^{-1}$, in words, that r_2 reverses the motion of r_1. But it is evident that this can happen only when r_1 and r_2 are rotations about the same axis. So the solution $(k,l,1)$ implies that there is only one axis. The solution $(k,1,1)$ says the same thing and it is perhaps simpler to use this one when we mean a one-axis polyhedron. This is the rotational symmetry of a pyramid.

Exactly as in the plane patterns, all the additional polyhedral groups can be derived from these by adding a mirror or the polyhedral analogue of glide, rotatory-reflection,* to extend the group.

Now let me give you drawings of a number of polyhedra. Try to classify them according to the solutions of the inequality $1/k + 1/l + 1/m > 1$,

* Rotatory-reflection is $r \cdot m$. The axis is perpendicular to the mirror plane; the direction of rotation is therefore parallel to the mirror.

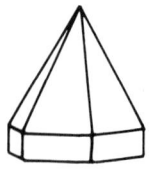 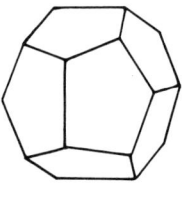 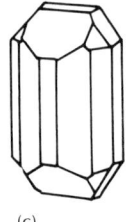  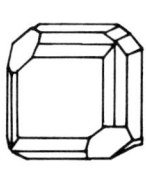 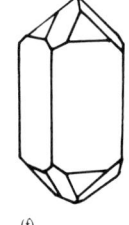

(a)　　　(b)　　　(c)　　　(d)　　　(e)　　　(f)

135 The Theory of Patterns

that is, according to their rotational symmetries. And then determine whether the symmetry group contains mirror reflection as well.

Festivant All right. And shall I bring this assignment to you this afternoon?

Symmetrica No, tomorrow morning. Afternoon sessions are strongly discouraged here. We prefer to hike, play chess, sing canons, or rest. It gives our thoughts a chance to crystallize. And this evening we have our weekly concert. A renowned harpsichordist is coming to play Bach's "Art of Fugue." We never tire of that.

Festivant I noticed the beautiful harpsichord in the library. How did you get it up here?

Symmetrica Pierre built it for us, on cloudy days.

IV. Subgroups and Subtle Symmetries
Time: the next morning
Place: the study

Festivant Here, on this sheet of paper, are my answers to the problems you assigned me.*

Symmetrica Festivant, this is really excellent. You have grasped the essentials of classification very quickly. Now we are ready to consider another question: what is the purpose of this classification?

Festivant I hope that it will shed light on relationships among the polyhedra.

Symmetrica Yes, if classification is to provide insight, it must be

---

* (a) (6,1,1), mirror symmetry; (b) (3,3,2), mirror symmetry; (c) (2,2,2), mirror symmetry; (d) (4,3,2), mirror symmetry; (e) (3,3,2), no mirror symmetry; (f) (4,2,2), mirror symmetry; (g) (4,3,2), mirror symmetry. Same group as (d); (h) (3,3,2), no mirror symmetry. Same group as (e); (i) (6,2,2), mirror symmetry; (j) (4,3,2), mirror symmetry. Same groups as (d), (g); (k) (2,2,2), mirror symmetry. Same group as c!

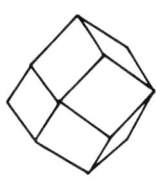
(g)
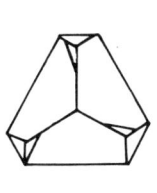
(h)
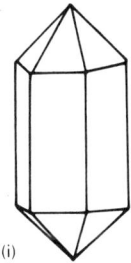
(i)
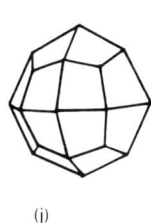
(j)
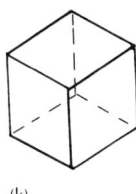
(k)

because there is a hierarchy of symmetries which helps us to find patterns within patterns. I will show you what I mean by an example. Draw a hexagon and decorate it to illustrate its symmetries. Which motifs illustrate only the rotational symmetries of the hexagon?

Festivant The motifs fall into two sets, each of which illustrates the rotational symmetries; one set is oriented one way, the other, the opposite way.

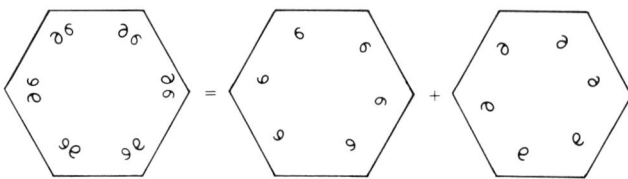

Symmetrica Yes, and in just the same way you can find two tetrahedra in a cube:

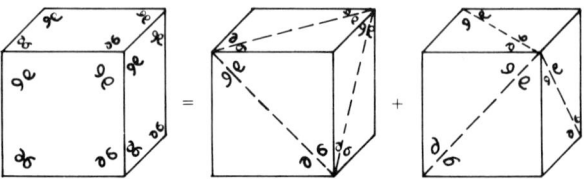

You see then that symmetry groups can contain smaller groups, subgroups, within them. Among the various classes of polyhedra there are many interesting group-subgroup relationships, systems within systems. A knowledge of the subgroups of a given group helps us to understand its structure. For example, we notice that the two oriented hexagons you found are related by reflection in a mirror, which we can denote by m. The symmetries of an oriented hexagon, all rotations, are i, r, r^2, r^3, r^4, r^5, $r^6 = i$. And if you check the products $i \cdot m$, $r \cdot m$, $r^2 \cdot m$, and so forth, you will see that these operations correspond to reflections in the six mirrors of the complete hexagon! This new set of operations is called a "coset" of the rotation group.

Festivant And can the symmetry group of the cube be described as the tetrahedral subgroup plus a coset?

Symmetrica Yes, it can. Now if you'll bear with me, I'll show you that the notion of a coset reveals some very interesting things about finite groups, that is, groups with a finite number of operations. Let us suppose that we are given a group, which we will call G, and any subgroup, which we will call H (most groups have many subgroups). Let us suppose that G contains n operations (including i) and that H contains k operations (again, including i). To begin, we list the k operations in H: $\{i, h_1, h_2, \ldots, h_{k-1}\}$. Now choose one of the operations in G which are not in H—suppose it is g_1—and form the k products $i \cdot g_1$, $h_1 \cdot g_1$, $h_2 \cdot g_1, \ldots, h_{k-1} \cdot g_1$. This is what we just did with the hexagon, but now we are talking more abstractly; G is any group, H any of its subgroups. For short, we denote this coset by Hg_1. Now, the first thing to note is that no operation in Hg_1 can be in H.

Festivant Why not?

Symmetrica I will show you that if one of them were in H, we would be forced to conclude that g_1 itself was in H, although we had expressly chosen it because it was not. For example, suppose that the product $h_1 \cdot g_1$ were in H. Now the closure and inverse properties of groups allow us to conclude that the product $h_1^{-1} \cdot h_1 \cdot g_1$ is also in H, but this expression can be simplified to $i \cdot g_i$, or more simply g_1. And that is our contradiction.

Festivant I see; that is a very clear and simple argument.

Symmetrica A similar argument will show that the k products in the coset are always different symmetry operations. So we see that if there is one operation in G which is not in H, then there are at least k of them which are not in H.

Now if the group is large and the subgroup small, the subgroup may have two or more distinct cosets. For example, after the first coset is formed, it is possible that there are some operations in G which are in neither H nor Hg_1. If that is so, we can form another coset, Hg_2, where g_2 is an operation not yet listed. One can prove that the k operations in Hg_2 will be different from one another as well as from all those in H and in the coset Hg_1. That means that G has at least k more operations, that is, at least $3k$. And, after this, if G has any more left over we can form a third coset, then a fourth, and so on. Eventually there won't be any operations in G left over. Then we find that G has been partitioned into a number, say m, of sets of k operations each: H, Hg_1, Hg_2, and so forth.

Festivant Then the number of operations in a subgroup always is a divisor of the number of operations in the group.

Symmetrica That's right! You've discovered "Lagrange's Theorem," an important fact about finite groups. You see, you have climbed a little further up our mountain.

Festivant Are cosets, then, a key to the study of symmetry?

Symmetrica Cosets are perhaps the *sine qua non*. We can build groups by forming cosets—that's how we add reflections to the rotation groups of plane patterns and polyhedra. And you will find that cosets reveal hidden patterns. For example, cosets provide the clue to understanding change-ringing compositions.

Festivant I would very much like to hear about that. I enjoyed the essay by Alice Dickinson, and I hope I will be able to hear your bells while I am here.

Symmetrica I believe some of our students will be ringing this evening, so this is a good time for us to discuss it. You recall that, in a plain hunt on six bells, the changes in the orders in which the bells are rung are effected by just two permutations. The first, which we will call *a*, consists of transposing the first and second bells, the third and fourth, and the fifth and sixth. The second permutation, *b*, keeps the first and sixth bells in their places, and transposes the second and third, and the fourth and fifth.

Festivant I see that one goes from row to row in the table for Plain Bob Minor (p. 47) by performing first *a*, then *b*, then *a*, then *b* again, and so forth.

Symmetrica That's right. They can be combined in the same way the symmetry operations can—the symbol "·" again means "followed by." The permutations *a* and *b* generate a group (call it P) of permutations. P contains twelve different permutations since $a^2 = b^2 = i$; they are: i, a, $b \cdot a$, $a \cdot b \cdot a$, $(b \cdot a)^2$, $a \cdot (b \cdot a)^2$, $(b \cdot a)^3$, and so forth; $(b \cdot a)^6 = i$. This means that after ringing the bells in the position corresponding to $a \cdot (b \cdot a)^5$, if we were to apply *b* again, the composition would be over.

Festivant So that is why the ringers "dodge," that is, they perform an unusual permutation, which is neither *a* nor *b*.

Symmetrica Yes, the new permutation is one that is permitted by the rules of change ringing and which arranges the bells in a new order so that *a* and *b* can be repeated again without causing any repetition in the ringing. Look at the last entry in the first column of the table. The bells are rung in the order 135264. Now this order can be obtained from the original one (123456) by the permutation $c \cdot b$, where *c* keeps the first two bells fixed and transposes the third and fourth, and the fifth and sixth.

```
1 2 3 4 5 6
              via b
1 3 2 5 4 6
              via c
1 3 5 2 6 4
```

Dodging is analogous to finding a new symmetry operation, one not in the subgroup H. By dodging, one generates a coset of P and thus turns P into a subgroup of a larger group.

In the second column, you can see that these permutations form a coset of P, $P(c \cdot b)$. At the end of the second column, the same dodge is performed again to avoid repeating 135264. So the third column represents the coset $P(c \cdot b)^2$, and the fourth represent $P(c \cdot b)^3$. If you listen carefully this evening, you can hear the cosets!

Festivant Why are there only five columns?

Symmetrica Because $(c \cdot b)^5 = i$. So after repeating this dodge five times, we return to the original order. That is, we can't generate any more new permutations with it.

Festivant Could the ringing be continued if, at the end of the fifth column, the ringers employed a different dodge?

Symmetrica Yes, it could. This would amount to adding a new generator to the group, and so the process of forming cosets could be continued.

Festivant This is very exciting!

Symmetrica Oh, yes—and cosets play a key role in so many patterns, for example, in color symmetry. And the relation between the atomic pattern of a crystal and its external symmetries is also described by cosets. They have many theoretical applications too. I hope you will study them after you leave here. Shall we climb the tower this evening, at dusk? I think you will appreciate the view now.

Festivant Yes, I would like to very much. I must leave early tomorrow, and I am eager to see the view before I go.

V. Epilogue

Place: the tower

Time: early evening

Festivant At last we begin to climb the tower. How I have longed for this! I see that several others are here already. Who are they?

Symmetrica Some of my colleagues. Shh! We must not disturb them; they are deep in contemplation.

Festivant But I hope we may talk a little when we reach the top. At dinner, you gave me a list of books* that I might read when I have returned home. It's a long list, and I don't know where to start. Could you give me some suggestions?

Symmetrica Yes, I'll be glad to tell you which ones seem to be especially suited to a beginner. But I hope it won't cause you to neglect the others.

Festivant No, certainly not.

Symmetrica One book I heartily recommend is Tabourot's (or Arbeau's)

---

* This is the bibliography, pp. 150–153.

Orchesographie. This book is written as a dialogue between a dancing master and a young man who wishes to learn the dances of his time, the late sixteenth century. As you know, the dances then were literally patterns of symmetry. And in addition to his entertaining discourse, Arbeau provides the music to many of the dances, and delightful illustrations.

Festivant   I saw the Cambridge Court Dancers at the Symmetry Festival. They performed several of Arbeau's dances: Pavane and Tourdion, Branle des Pois, and Branle de la Haye, to name only a few.

Symmetrica   Did you attend the performance of Bach's *Musical Offering* too?

Festivant   Yes, I did.

Symmetrica   Then I would also suggest that you read *J. S. Bach's Musical Offering* by Hans David. This little book describes the symmetries of that elaborate composition in fascinating detail. For example, he shows that one of the canons (*Canon a 2 per tonus,* or the *Modulating Canon*) is a complex musical helix. The symmetries of the other canons, and of the entire work, are discussed as well.

Festivant   It fascinates me to think of helices in crystals, canons, staircases, and plants—the same architecture appearing in so many guises.

Symmetrica   Yes, and that brings a delightful book to mind, *The Architecture of Molecules,* by Linus Pauling and R. Hayward. The illustrations are magnificent, and the text elementary and concise. Another interesting book is the brief volume, *Researches on Molecular Asymmetry,* by Louis Pasteur. Pasteur was the first to deduce the relation between the structure of molecules and the presence or absence of mirror symmetry in the crystals they form. This discovery, made twenty-five years before the spatial arrangement of atoms and molecules was recognized to be as important as their numbers and kinds, paved the way for our present understanding of molecular and crystal architecture. Pasteur described his researches very lucidly in the two lectures which comprise this book.

Festivant   It appears that asymmetry can be as interesting as symmetry.

Symmetrica   I would agree, and suggest that you also read *The Ambidextrous Universe,* by Martin Gardner, in this connection.

Festivant   I think I would like to learn more about the consequences of asymmetry. It's intriguing to think that it could matter.

Symmetrica   Then if you have some kitchen space, you should try to grow some crystals. In *Crystals and Crystal Growing,* by Alan Holden and Phyllis Singer, there are some "recipes" for growing crystals and suggestions for interesting experiments. The physical manifestations of asymmetry in sodium chlorate and rochelle salt are particularly striking, and these two crystals are easy to grow.

141   The Theory of Patterns

Festivant   I tried to grow sugar crystals once, but it was a mess.

Symmetrica   These two are much easier to grow than sugar crystals! Now, returning to symmetry, another book you should not miss is *The Grammar of Ornament,* by Owen Jones. This enormous and beautifully illustrated collection of ornaments from all times and places was published in 1910. Of course, the architect of the Belvedere Center, M. C. Escher, is without peer in modern symmetrical design, and you will enjoy *The Graphic Work of M. C. Escher.* And as a companion volume, you must read *Symmetry Aspects of M. C. Escher's Periodic Drawings,* by Caroline MacGillavry. In it, she analyzes the symmetries in many of Escher's drawings, including the color symmetries, and describes many analogies between these patterns and problems in crystal structure. Perhaps *Regular Figures,* by L. Fejes Toth, will be of interest to you too. This book is an expository discussion of the geometry of patterns and packing problems. It's fairly elementary. A more advanced mathematics text, which treats geometrical symmetry from the standpoint of group theory and thus continues the study you've begun here, is *Geometry and Symmetry,* by Paul Yale. It's not elementary, but it's worth the effort it takes to read it.

Festivant   Are there any books which deal with structures in space?

Symmetrica   Yes, there are two on the list: *Space Structures,* by Arthur Loeb, and *Shapes, Space, and Symmetry,* by Alan Holden. Both of these are introductory. And if you would like to explore shapes in four or more dimensions, read *Regular Polytopes,* by H. S. M. Coxeter.

Festivant   Thank you! I think this will give me a good start.

Symmetrica   But, I must mention two more. One is *On Growth and Form,* by D'Arcy W. Thompson. In this book physical and biological structures of many kinds, from snowflakes to fish, are described, and their analogues and interrelationships discussed. The book is written in a beautiful literary, discursive style. Also you must read *Symmetry,* by Hermann Weyl. This is also a work of erudition, a masterpiece of composition. It surveys the entire scope of symmetry from a majestic height, and it is always a source of inspiration to us here at Belvedere. We are deeply indebted to both Thompson and Weyl.

Festivant   And I am indebted to you, Symmetrica, for showing me this vast perspective.

(The tower bells begin to ring)
The End

GEORGE FLECK **A Final Focus**

In "Mending Wall," Robert Frost writes of a New England neighbor, on the other side of their common stone fence, who lives by memorized maxims. Mr. Frost observes that because the neighbor "will not go behind his father's saying that 'Good fences make good neighbors', he moves in darkness." To make wise use of the saying, one needs to know why (and especially when) good fences make good neighbors.

Scholars have their memorized maxims, too. Succinct summaries (call them fundamental principles, basic postulates, scientific laws) are certainly convenient in the classroom. It is tempting when studying the abstract, terse literature of science and mathematics to accept and to memorize many pithy summary sayings. But then, once memorized, the sayings may seem like literal, revealed truths. And strange truths at that, for often the saying without a full context is rather empty of meaning. So we often need to go behind the conventional textbook sayings if we are to know why (and when) the sayings are so. Otherwise, we may unwittingly move in darkness.

Going behind the saying may not be a straightforward task. There is a deceptive appearance of wholeness, completeness, and separateness to most of the intellectual systems of mathematics and the physical sciences. The essentials are all in place, looking neat and tidy. Everything seems valid in its own particular terms. But what about meaning, significance, and truth? Meaning and significance require connections to be made with a larger world, usually left to be supplied by the reader. Truth remains ever elusive. We could often use some fruitful strategies for going behind a physical law or a mathematical formalism. Is there a way out from inside, a way out and around?

One often-recommended strategy is to explore in depth an application of the formalism, probably doing this research with familiar subject matter, conducting the exploration intensively and exhaustively. This is the way of the specialist. Another strategy, very different from the first and likely to yield different insights, is to compare and contrast the ways in which various specialists apply the formalism in various situations, with various subject matter, seeking various goals. This is the way of the generalist. Of course, the same person may be a specialist one day, a generalist the next.

Because specialists in quite diverse fields could be persuaded to share with us their work with symmetry and their ideas about symmetry, we were able to make *Patterns of Symmetry* a sourcebook for generalizing. We have deliberately restrained our own impulses for explicit generalizing, but instead have assembled and juxtaposed material to provoke comparisons, to suggest connections, and to stimulate you to draw your own tentative generalizations. Your unique background and your own personal perspectives will, no doubt, influence your conclusions. In any case, we hope that you will have the courage to go behind the sayings about symmetry in your fields of particular interest.

There is in our midst a far-reaching intellectual world of symmetry, but it is seldom seen as a whole. We have sought to give tangible expression, in a faithful, realistic, and concrete manner, to that world. Here the geology student studying mineralogy, the chemistry student studying molecular structure, and the mathematics student studying group theory are members together of a larger community that includes dancers, musicians, printers, writers, weavers, painters, architects, and bakers. Hopefully, we have helped some chemists and musicians to recognize their kinship. We have assembled opportunities for specialists, would-be specialists, and generalists to listen to, to see the work of, and to work with others whose interests are linked by relationship to symmetry concepts. This world of symmetry is a community of people: amateurs and professionals, hobbyists and craftsmen, students and teachers. It is also a collection of ideas and systems (some carefully codified, others vague and elusive), and many tangible things (paintings, crystals, polyhedral models, and symmetrical cookies, to suggest just a sampling). This larger world is part of the context for the sayings of specialized symmetry study.

Becoming an expert—a journeyman printer or typographer, a skilled musician or composer, a qualified crystallographer or mathematician—requires a disciplined apprenticeship (a discipleship, as it were), and a continuing disciplined approach to one's work. And so the word *discipline* is often used to describe a field such as typography, music, or mathematics. Staying, studying, and working securely within a discipline seems a safe way of life, and it oftentimes is a productive way of life. An established discipline has its own subject matter, its own procedures, its own standards, and its own rewards. It may seem to be a neat parcel of knowledge and a well-defined area of inquiry, conveniently walled off from the rest of the world. And yet, as Mr. Frost observed when commenting on his stone wall, "Something there is that doesn't love a wall/That wants it down." The conventional dividing walls between disciplines are often arbitrary, and should not be taken too seriously when they block interesting paths. There often is no need to limit oneself to just one discipline (at least not just one, chosen at age eighteen, persisted in forever), nor necessarily to wear the categorizing name-tag of any discipline. Sometimes different name-tags imply subject matter boundaries and walls where boundaries need not and ought not be. "Before I build a wall I'd ask to know," says Robert Frost, "what I was walling in or walling out."

Symmetry study as a whole benefits from interdisciplinary scholarship and from the work of generalists. There are also opportunities for significant contributions from specialists, and for much important single-discipline research. How does an individual find a place in such a field? Are there some people who should be specialists and others who should be general practitioners? Or is there, for each person, a time for specializing and a time for generalizing? These questions are important, but the answers are not easy. To make an informed choice of a specialty, one needs a reliable general overview of the field. To make a contribution as a responsible generalist, one needs at least moderate competence in several specialties. A certain degree of maturity and wisdom is needed in making these choices. These questions were raised at the 1973 Smith College Symmetry Festival and again in a discussion at Smith College on May 1, 1974*. The following excerpts are from a recording of the 1974 discussion:

Lory Wallfisch We knew, very well, one of the greatest architects, Frank Lloyd Wright. We lived on one of his estates for about five months providing music for him, his family, and his apprentices. Now this great artist, who was then nearly eighty years old and world famous, had apprentices—and mind you that was the designation. It was not a school to give out degrees or diplomas, yet young people who wanted to become architects came from all over the world and lived near him and learned from him to become architects and to assimilate his principles and his views. These fifty young would-be architects were supposed to do everything from planting corn, in other words working in the fields on his estate in Wisconsin, peeling potatoes, and running the household—cooking, doing the housework, milking cows (there were forty cows), producing the butter, making the homemade ice cream—running everything, including the masonry repairs on the buildings, and most of them only got to the drafting board late at night.

George Fleck This is the idea of being an amateur, and learning something about how the world works, as well as being a theoretician.

---

* The participants included Professor and Mrs. Arthur Loeb, three members of the Smith faculty, Mary Murphy '75 (all involved in the Symmetry Festival), and undergraduate students from the courses Symmetry, Mineralogy, and Structure of Molecules.

Lory Wallfisch Yes, the word amateur is a very interesting one. It does not limit the degree of specialization. In fact, I have known many amateur violinists who are better than the so-called professional. Actually, I think that the dictionary definition is, "an amateur is that one who does not earn his living. . . ." I had enormous respect for those young architects, many who have become very successful. They left after two years, having learned all the things I mentioned and having been exposed to music—that's how we got there, as court musicians. Mr. Wright insisted that Sunday dinner was to be followed by a live concert, chamber music mostly, in his private theatre—which was a dream of a little theatre. He had art movies shown also. Then we started to precede the Saturday night movie by a half hour concert, and he decided it was a long way from Sunday until the next Saturday and so we played a small concert on Wednesday evenings, all for these young architects. They got up at 6:30 in the morning—we were all at breakfast (Mr. Wright included) at 7:30—and were supposed to sing. (They weren't only passively exposed.) Palestrina, whatever, the best of music. His secretary conducted the choir every morning for half an hour.

George Fleck That's the myth of small-town America, too. I think about the colonial "architects" who built New England. They were common, ordinary builders who also sang in the church choir and milked the cows and ran the politics and everything else.

Lory Wallfisch One word that hasn't been used at all, in connection with the word amateur, is "hobby." The hobby is what enriches. You are a specialist, but having one or more hobbies is what leads your curiosity and interest to one thing or another. That's one nice aspect of American life, I think. You are asked in this country, "What's your hobby?" more than in Europe.

Lotje Loeb Yes, and it's looked at with respect

Lory Wallfisch It even appears on some written questionnaires. I think this is very laudable.

Arthur Loeb But somehow it becomes competitive. . . . In Renaissance dancing, there should be no competition, because everyone has to cooperate. But even here you find that sort of thing. The Court Dancers are traveling more and others start doing it, and then the leaders want competitive teams. Here's the culture pushing the things out of context. To what extent is this still authentic recreation?

Lory Wallfisch To be well-versed in many things, you have to have that courage, and sense of adventure—I understood what you meant about the risky decision—to have faith in what you pursue.

Arthur Loeb It also means, I think, a fair amount of curiosity. There are people who say there are certain critical moments in one's life that determine the remainder of the future, and if you make the decision one way or another that could ruin your life—I don't go along with that at all. There are such decision points but you usually have a great many of them, if you're open. If you are active, you find that these things happen often. When you look back, you may find that eighty percent of them did not turn out to be critical decision points and twenty percent of them did. If you happen to miss those, there would have been others that would have led to exactly the same road. It's like the redundancy in systems design.

Bernadette Margel Suppose a person investigates and his curiosity leads him in one particular vein, and he wants to specialize. I don't think we should move the pendulum so far to the other side that we look down on the person who wants to specialize. I don't think we should impose the standard of a Renaissance Man on everybody. I think people should be allowed to opt.

Marjorie Senechal Something that Mrs. Wallfisch said is appropriate here. For different people the need for being specialized comes at different times. For some people there is a great need to specialize for awhile, because you really want to know a lot about one thing, and then you want to have something else to add to that, you want to broaden out and see how it relates to other things, and then later you want to know more about one thing again—it may not be the same thing. It alternates back and forth, between specialization and generalization, and if you know the value of both, then you're not afraid to do either one.

We hope that your imagination has been stimulated while reading and looking at *Patterns of Symmetry*. You may find yourself looking for symmetry elements in the most unlikely of places. Symmetry study still has much to contribute to several of the academic disciplines. And many fields have much to give to symmetry studies. We hope that you will find creative opportunities (as an amateur or as a professional) to make some of these contributions.

Selected Bibliography

The subject of symmetry is a vast one. Conceived as design or pattern in the arrangement of matter or thought, it embraces all the sciences and the arts, and is a reflection of man's search for order and unity in the universe. Therefore this bibliography can be only a very small selection from the rich treasury of works on the subject, with inevitable omissions, and is intended primarily to indicate the wide scope of the subject.

This selected bibliography was compiled by the Reference Department of the William Allan Neilson Library, Smith College, for the Vanderbilt Symmetry Festival, February 1973, and revised for *Patterns of Symmetry* by the editors.

Albers, Josef. *Homage to the Square.* Zurich: Gimpel und Hanover Galerie, 1965.

Alighieri, Dante. *The Comedy of Dante Alighieri, the Florentine.* Translated by Dorothy L. Sayers. New York: Basic Books, 1962; New York: Penguin, 1949.

Azároff, Leonid V. *Introduction to Solids.* New York: McGraw-Hill, 1960.

Bentley, W. A., and Humphreys, W. J. *Snow Crystals.* New York and London: McGraw-Hill, 1931; New York: Dover, 1962.

Bernal, Ivan, Hamilton, W. C. and Ricci, J. S. *Symmetry: A Stereoscopic Guide for Chemists.* San Francisco: W. H. Freeman, 1972.

Birss, Robert R. *Symmetry and Magnetism.* Amsterdam: North-Holland Publishing Co., 1964.

Bloss, Fred D. *Crystallography and Crystal Chemistry.* New York: Holt, Rinehart and Winston, 1971.

Borges, Jorge Luis. *Ficciones.* Edited by A. Kerrigan. New York: Grove Press, 1962.

Bridge, Sir J. F. *Double Counterpoint and Canon.* London: Novello, 1881.

Brieger, Peter H. *Illuminated Manuscripts of the Divine Comedy.* Princeton: Princeton University Press, 1969.

Bucher, François. *Joseph Albers: Despite Straight Lines.* New Haven: Yale University Press, 1961.

Correns, Carl W. *Introduction to Mineralogy, Crystallography, and Petrology.* 2d ed. Translated by W. D. Johns. New York: Springer-Verlag, 1969.

Cotton, F. A. *Chemical Applications of Group Theory.* 2d ed. New York: Wiley-Interscience, 1971.

Coxeter, H. S. M. *Regular Polytopes.* New York: Macmillan, 1963; New York: Dover, 1973.

Coxeter, H. S. M., and Moser, W. O. J. *Generators and Relations for Discrete Groups.* Berlin: Springer Verlag, 1957.

David, Hans. *J. S. Bach's Musical Offering.* New York: G. Schirmer, Inc., 1945.

Dehio, G. "Zur geschichte der Buchstabenreform in der Renaissance." *Repertorium fuer Kunstwissenschaft* 4 (1881), 269–279.

Dürer, Albrecht. *Of the Just Shaping of Letters.* Translated by R. T. Nichol from Latin text of 1535 ed. New York: Dover, 1965.

Escher, Maurits Cornelis. *The Graphic Work of M. C. Escher.* Translated by John E. Brigham. New York: Meredith Press, 1967; New York: Ballantine, 1971.

Fackler, John P., Jr. *Symmetry in Coordination Chemistry*. New York: Academic Press, 1971.

Fejes Tóth, L. *Regular Figures*. New York: Macmillan, 1964.

Feliciano, Felice. *Alphabetum Romanum*. Verona: Editiones Officinae Bodini, 1960.

Frankl, Paul. *The Gothic Literary Sources and Interpretations*. Princeton: Princeton University Press, 1960.

Gardner, Martin. *The Ambidextrous Universe*. New York: Basic Books, 1964.

Hall, Lowell H. *Group Theory and Symmetry in Chemistry*. New York: McGraw-Hill, 1969.

Herbert, George. *The Works of George Herbert*. Edited by F. E. Hutchinson. Oxford: The Clarendon Press, 1941.

Hilton, Harold. *Mathematical Crystallography and the Theory of Groups of Movements*. Oxford: Clarendon Press, 1903; New York: Dover Publications, 1963.

Hochstrasser, Robin M. *Molecular Aspects of Symmetry*. New York: W. A. Benjamin, 1966.

Hoff, Jacobus H. van 't. *The Arrangement of Atoms in Space*. 2d rev. ed. Translated with a preface by Johannes Wislicenus. Appendix, "Stereochemistry among Inorganic Substances," by Alfred Werner. London: Longmans, Green & Co., 1898.

Holden, Alan. *Shapes, Space, and Symmetry*. New York: Columbia University Press, 1971.

Holden, A., and Singer, P. *Crystals and Crystal Growing*. Garden City, New York: Doubleday, Anchor, 1960.

Hollas, J. Michael. *Symmetry in Molecules*. London: Chapman and Hall Ltd., 1972.

Humphrey, Doris. *The Art of Making Dances*. Edited by Barbara Pollack. New York: Rinehart 1959; New York: Grove Press, Evergreen, 1959.

Hurlbut, Cornelius S. *Minerals and Man*. New York: Random House, 1968.

Jaeger, Frans Maurits. *Lectures on the Principle of Symmetry and its Applications in all Natural Sciences*. Amsterdam: Publishing Company "Elsevier," 1917.

Jaffé, Hans H., and Orchin, M. *Symmetry in Chemistry*. New York: John Wiley, 1965.

Jones, Owen. *The Grammar of Ornament*. London: Bernard Quaritch, 1910.

Kepes, Gyorgy, ed. *Module, Proportion, Symmetry, Rhythm*. New York: G. Braziller, 1966.

Kramer, Etel Thea. *The Ornament of Louis Sullivan*. Northampton, Mass.: Smith College, 1960.

Laban, Juana de, ed. *Institute of Court Dances of the Renaissance and Baroque Periods*. New York, Committee on Research in Dance, Dance Notation Bureau, 1972.

Le Corbusier. *The Modulor*. Cambridge, Mass.: Harvard University Press, 1954. Cambridge, Mass.: M.I.T., 1968.

Le Corbusier. *Modulor 2, 1955*. Cambridge, Mass.: Harvard University Press, 1958; Cambridge, Mass.: M.I.T., 1968.

Lehr, Roland, E., and Marchand, A. P. *Orbital Symmetry: A Problem-Solving Approach*. New York: Academic Press, 1972.

Loeb, Arthur L. *Color and Symmetry*. New York: Wiley-Interscience, 1971.

Loeb, Arthur L. *Space Structures*. Reading, Mass.: Addison-Wesley, 1976.

MacGillavry, Caroline. *Symmetry Aspects of M. C. Escher's Periodic Drawings.* Utrecht: A. Oosthoek's Ultgeversmaatschappij, 1965.

Mann, Alfred. *The Study of Fugue.* New Brunswick, N.J.: Rutgers University Press, 1958.

Nakaya, Ukichiro. *Snow Crystals: Natural and Artificial.* Cambridge, Mass.: Harvard University Press, 1954.

Offner, Eliot. *The Granjon Arabesque.* Northampton, Mass.: Rosemary Press, 1969.

Pasteur, Louis. *Researches on the Molecular Asymmetry of Natural Organic Products.* 1860. Alembic Club Reprints, no. 14. Edinburgh: E. and S. Livingstone Ltd., 1948.

Pauling, Linus, and Hayward, R. *The Architecture of Molecules.* San Francisco: W. H. Freeman, 1964.

Poe, Edgar Allan. "Eureka, a Prose Poem." *Complete Works of Edgar Allan Poe.* Edited by James Harrison. vol. 16. New York: Fred de Fau, 1902.

Ryde, John. *A Suite of Fleurons, or a Preliminary Enquiry into the History and Combinable Natures of Certain Printers' Flowers.* Boston: C. T. Brandford, 1957.

Sabbe, Maurits and Andin, Marius. *Die Civilité-Schriften des Robert Granjon.* Wien: Bibliotheca Typographica, 1929.

Shubnikov, A. V. and Koptsik, V. A. *Symmetry in Science and Art.* Translated by G. D. Archard. Edited by David Harker. New York: Plenum Press, 1974.

Simon, Hilda. *The Splendor of Iridescence; Structural Colors in the Animal World.* New York: Dodd, Mead, 1971.

Sinnott, E. W. *Plant Morphogenesis.* New York: McGraw Hill, 1960.

Spies, Werner. *Albers.* New York: H. N. Abrams, 1970.

Strang, Gerald. "Mirrorrorrim." *New Music* 5, no. 4 (July 1932).

Sullivan, Louis H. *A System of Architectural Ornament, According with a Philosophy of Man's Powers.* New York: Press of the American Institute of Architects, Inc., 1924.

Tabourot, Jehan [Thoinot Arbeau]. *Orchesographie.* Paris: F. Vieweg, 1888; Translated by Mary Evans. Edited by Julia Sutton. New York: Dover, 1966.

Thompson, D'Arcy Wentworth. *On Growth and Form.* Cambridge, England: Cambridge University Press, 1917; New York: Macmillan, 1942.

Torniello, Francesco. *The Alphabet of Francesco Torniello.* Verona: Editiones Officinae Bodini, 1971.

Tory, Geoffroy. *Champ Fleury (1529).* Translated and annotated by G. B. Ives. Illustrated by Bruce Rogers. New York: The Grolier Club, 1927; New York: Dover, 1967.

Tovey, Donald Francis. *A Companion to "The Art of Fugue" of J. S. Bach.* London: Oxford University Press, 1931.

Urch, D. S. *Orbitals and Symmetry.* Harmondsworth: Penguin paperback, 1970.

Vitali, Giovanni Battista. *Artifici Musicali; opus XIII.* Edited by Louise Rood and Gertrude Smith. Northampton, Mass.: Smith College, 1959.

Vitruvius. *Ten Books on Architecture.* Translated by Morris Hickey Morgan. New York: Dover, 1960.

Wellek, René, and Warren, Austin. *Theory of Literature.* New York: Harcourt, Brace, 1949.

Wells, Alexander Frank. *Models in Structural Inorganic Chemistry.* New York: Oxford University Press, 1970.

Westinghouse Research Laboratories. *Crystals Perfect and Imperfect*. New York: Walker, 1965.
Weyl, Hermann. *Symmetry*. Princeton: Princeton University Press, 1952.
Wilson, Wilfrid G. *Change Ringing: The Art and Science*. New York: October House, 1965.
Wittkower, Rudolf. *Architectural Principles in the Age of Humanism*. London: Alec Tiranti, 1952; New York: Random House, 1965.
Woodward, R. B., and Hoffman, Roald. *The Conservation of Orbital Symmetry*. Weinheim: Verlag Chemie, 1970.
Wrinch, Dorothy Maud. *Chemical Aspects of Polypeptide Chain Structures and the Cyclol Theory*. Copenhagen: Munksgaard, 1965.
Yale, Paul B. *Geometry and Symmetry*. San Francisco: Holden-Day, 1968.

Notes

Introduction
1. Weyl, H., *Symmetry* (Princeton: Princeton University Press, 1952).
2. Brewster, D., *A Treatise on the Kaleidoscope* (Edinburgh: A. Constable, 1819).

The Many Facets of Symmetry
1. Guglielmo Ebreo, ca. 1460 (Siena: Biblioteca Communale degli Intronati, Codex L.V. 29).
2. Domenico da Piacenza, ca. 1420 (Paris: Bibliothèque Nationale, fds. it. 972).
3. Guglielmo Ebreo, ca. 1460 (Paris: Bibliothèque Nationale, fds. it. 972).
4. Domenico da Piacenza, ca. 1420.

Color Symmetry and Its Significance for Science
1. MacGillavry, C. H., *Symmetry Aspects of M. C. Escher's Periodic Drawings* (Utrecht: A.-Oosthoek, 1965).
2. Loeb, A. L., *Color and Symmetry* (New York: John Wiley & Sons, 1971).

Change Ringing: Theory and Practice
1. C. A. W. Troyte, *Change Ringing: An Introduction to the Early Stages of the Art of Church or Hand Bell Ringing For the Use of Beginners*. 4th edition. London: Simpkin, Marshall, Hamilton, Kent & Co., Ltd., 1882, p. 160.

Renaissance Typographic Ornament
1. Smith College owns two Ratdolt books which are kept in the Rare Book Room of the William Allan Neilson Library.
2. Frantz, M. Alison, "Byzantine Illuminated Ornament," *Art Bulletin,* vol. 16 (1934), pp. 43-76.
3. Mardersteig's edition of 160 copies was already spoken for and thus out of print before publication. One copy was given to Smith College and is in the Rare Book Room of the William Allan Neilson Library.
4. Potter, Esther, "Giovanni Antonio Tagliente ca. 1465-1527," in *Splendour of Ornament* by Stanley Morison (London: Lion and Unicorn Press at the Royal College of Art, 1968).

Ambiguities in Symmetry-Seeking
1. Eliot, T. S., "Reflections on Vers Libre," *To Criticize the Critic,* New York, Farrar, Straus & Giroux, 1965, p. 185.

2. Whitman, C. H., "The Geometric Structure of the *Iliad*," in *Homer and the Heroic Tradition*, New York: W. W. Norton & Co., 1965.
3. Borges, J. L., "Death and the Compass," *Ficciones*, ed. A. Kerrigan, New York: Grove Press 1962, p. 129. All further quotations from Borges' stories, except where noted, are from this volume.
4. Borges, J. L., "The Draped Mirrors," *Dreamtigers*, New York: E. P. Dutton & Co., 1970, p. 27.
5. *Ibid.*, "Mutations," p. 41; "On Beginning the Study of Anglo-Saxon Grammar," p. 85.
6. Aeschylus, *Agamemnon*, trans. R. Lattimore, Chicago: University of Chicago Press, pp. 46, 41.
7. *The World of M. C. Escher*, ed. J. L. Locher, New York: H. N. Abrams, 1971, fig. 211. Escher's concern for periodic and symmetric elements in his graphics does not invariably lead to the sense of closure we get in "Swans." Sometimes, as in "Night and Day," the effect is more open and playful, the pattern-making more a moving beyond boundaries.
8. Wheelis, Allen, "How People Change," *Commentary* (May 1969), pp. 50–61.
9. Kafka, Franz, "Reflections on Sin," *The Great Wall of China*, trans. W. & E. Muir (London, 1933), p. 261.
10. Tucker, G. J. and Whitbeck, C., "Thought Disorder: Implications of a New Paradigm." Presented at the 125th Annual Meeting of the American Psychiatric Association at Dallas, Texas in May 1972.
11. Halprin, L., *The RSVP Cycles: Creative Processes in the Human Environment*, New York: George Braziller, Inc., 1969. This study of "scoring" environments and events is skillful in subverting symmetry.
12. Mann, T., *The Magic Mountain*, trans. H. T. Lower-Porter, New York: Random House, 1969, p. 480.
13. Globus, Gordon G., "Unexpected Symmetries in the 'World Knot'," *Science*, vol. 180, no. 4091 (June 15, 1973), 1129–1136.

Library of Congress Cataloging in Publication Data

Main entry under title:

Patterns of symmetry.

Includes the Proceedings of the Symmetry Festival, held at Smith College, February 1973.

Bibliography: p.

1. Proportion (Art)—Addresses, essays, lectures.
2. Arts—Addresses, essays, lectures.   I. Senechal, Marjorie.   II. Fleck, George M.   III. Symmetry Festival, Smith College, 1973.
NX650.P65P37   700'.1   76-56775
ISBN 0–87023–345–9